Modern Organic Chemistry Series

KENNETH L. RINEHART, JR., Editor

Volumes published or in preparation

N. L. ALLINGER and J. ALLINGER	**STRUCTURES OF ORGANIC MOLECULES**
TRAHANOVSKY	**FUNCTIONAL GROUPS IN ORGANIC COMPOUNDS**
STEWART	**THE INVESTIGATION OF ORGANIC REACTIONS**
SAUNDERS	**IONIC ALIPHATIC REACTIONS**
GUTSCHE	**CHEMISTRY OF CARBONYL COMPOUNDS**
PRYOR	**INTRODUCTION TO FREE RADICAL CHEMISTRY**
STOCK	**AROMATIC SUBSTITUTION REACTIONS**
RINEHART	**OXIDATION AND REDUCTION OF ORGANIC COMPOUNDS**
DePUY and CHAPMAN	**MOLECULAR REACTIONS AND PHOTOCHEMISTRY**
IRELAND	**ORGANIC SYNTHESIS**
DYER	**APPLICATIONS OF ABSORPTION SPECTROSCOPY OF ORGANIC COMPOUNDS**
BATES and SCHAEFER	**RESEARCH TECHNIQUES IN ORGANIC CHEMISTRY**
TAYLOR	**HETEROCYCLIC COMPOUNDS**
HILL	**COMPOUNDS OF NATURE**
BARKER	**ORGANIC CHEMISTRY OF BIOLOGICAL COMPOUNDS**
STILLE	**INDUSTRIAL ORGANIC CHEMISTRY**
SIM, RINEHART NORMAN, and GILBERT	**X-RAY CRYSTALLOGRAPHY, MASS SPECTROMETRY, AND ELECTRON SPIN RESONANCE OF ORGANIC COMPOUNDS**
BATTISTE	**NON-BENZENOID AROMATIC COMPOUNDS**

RESEARCH TECHNIQUES IN ORGANIC CHEMISTRY

Robert B. Bates

Department of Chemistry
University of Arizona

John P. Schaefer

Dean, College of Liberal Arts
University of Arizona

PRENTICE-HALL, INC., ENGLEWOOD

C—13-774489-7
P—13-774471-4

Library of Congress Catalog Card Number 74-140411
Printed in the United States of America

PRENTICE-HALL INTERNATIONAL, INC., London
PRENTICE-HALL OF AUSTRALIA, PTY. LTD., Sydney
PRENTICE-HALL OF CANADA, LTD., Toronto
PRENTICE-HALL OF INDIA PRIVATE LIMITED, New Delhi
PRENTICE-HALL OF JAPAN, INC., Tokyo

Current Printing (last digit):
10   9   8   7   6   5   4   3   2   1

To our wives

Foreword

Organic chemistry today is a rapidly changing subject whose almost frenetic activity is attested by the countless research papers appearing in established and new journals and by the proliferation of monographs and reviews on all aspects of the field. This expansion of knowledge poses pedagogical problems; it is difficult for a single organic chemist to be cognizant of developments over the whole field and probably no one or pair of chemists can honestly claim expertise or even competence in all the important areas of the subject.

Yet the same rapid expansion of knowledge—in theoretical organic chemistry, in stereochemistry, in reaction mechanisms, in complex organic structures, in the application of physical methods—provides a remarkable opportunity for the teacher of organic chemistry to present the subject as it really is, an active field of research in which new answers are currently being sought and found.

To take advantage of recent developments in organic chemistry and to provide an authoritative treatment of the subject at an undergraduate level, the *Foundations of Modern Organic Chemistry Series* has been established. The series consists of a number of short, authoritative books, each written at an elementary level but in depth by an organic chemistry teacher active in research and familiar with the subject of the volume. Most of the authors have published research papers in the fields on which they are writing. The books will present the topics according to current knowledge of the field, and individual volumes will be revised as often as necessary to take account of subsequent developments.

The basic organization of the series is according to reaction type, rather than along the more classical lines of compound class. The first ten volumes in the series constitute a core of the material covered in nearly every one-year organic chemistry course. Of these ten, the first three are a general introduction to organic chemistry and provide a background for the next six, which deal with specific types of reactions and may be covered in any order. Each of the reaction types is presented from an elementary viewpoint, but in a depth not possible in conventional textbooks. The teacher can decide how much of a volume to cover. The tenth examines the problem of organic synthesis, employing and tying together the reactions previously studied.

The remaining volumes provide for the enormous flexibility of the series. These cover topics which are important to students of organic chemistry and are sometimes treated in the first organic course, sometimes in an intermediate course. Some teachers will wish to cover a number of these books in the one-year course; others will wish to assign some of them as outside reading; a complete intermediate organic course could be based on the eight "topics" texts taken together.

The series approach to undergraduate organic chemistry offers then the considerable advantage of an authoritative treatment by teachers active in research, of frequent revision of the most active areas, of a treatment in depth of the most fundamental material, and of nearly complete flexibility in choice of topics to be covered. Individually the volumes of the Foundations of Modern Organic Chemistry provide introductions in depth to basic areas of organic chemistry; together they comprise a contemporary survey of organic chemistry at an undergraduate level.

KENNETH L. RINEHART, JR.

University of Illinois

Preface

This book is intended to serve as a guide for students who have completed at least two semesters of organic laboratory and are beginning organic research as advanced undergraduates or first year graduate students. In a short book on a broad topic, an attempt has been made to include many bits of information useful at this stage of expertise and to provide a reasonably complete survey of the laboratory techniques currently most important to organic chemists. The emphasis in presenting the techniques is on practical aspects. Thus, the scope of each method is discussed and, where expensive equipment is required, the approximate cost is given.

LABORATORY SAFETY

Organic research work very often involves flammable, poisonous, and explosive substances. Adherence to the following rules will help considerably to reduce the number and seriousness of accidents.†

Keep flammable solvents away from ignition sources. In the research laboratory, these sources include not only flames, but also *heating mantles, hot plates, electric stirring motors, cigarettes,* and *static sparks.* Even a steam bath can ignite an extremely flammable substance such as carbon disulfide.

Store flammable solvents safely. Keeping solvent containers in cabinets below bench tops rather than on or above them helps to keep fires small. Large amounts of solvents should be stored in metal containers, since large glass containers are relatively easily broken and the spilled solvent can be the cause of a very serious fire.

Wear safety glasses or regular glasses whenever you are in the laboratory. Safety lenses for regular glasses cost a bit more but are much less likely to shatter if hit by flying glass from an explosion.

Run reactions involving explosive substances behind safety shields. Portable plastic shields cost about $40. Hoods with panels which slide horizontally are also suitable. Manipulations are performed by *reaching around* the shield or panel while wearing asbestos gloves ($5 per pair).

† For further information, see the comprehensive work edited by N. V. Steere, *CRC Handbook of Laboratory Safety,* Chemical Rubber Company, 18901 Cranwood Parkway, Cleveland, Ohio 44128, which includes tables of hazard information for over 1000 chemicals. Also recommended are the booklets *Safety in Handling Hazardous Chemicals,* available from Matheson, Coleman, and Bell, Box 7203, Los Angeles, Calif. 90022, and *A Guide for Safety in the Chemistry Laboratory,* available from the Department of Chemistry, University of Illinois, Chicago, Ill. 60680.

Run reactions involving poisonous gases in hoods. The air flow in a hood should be tested occasionally with something light like tissue paper.

Strap gas cylinders securely, whether in use or in storage. Even if the gas it contains is not toxic or flammable, a cylinder can be dangerous. If it is knocked over, the valve may snap off, and if the pressure in the cylinder is high, it will be literally rocket-powered and can do considerable damage. Safety cylinder supports cost about $6.

LITERATURE

The most comprehensive work in English covering many aspects of laboratory techniques in detail is Weissberger's series of monographs *Technique of Organic Chemistry.*† These should be consulted whenever a detailed discussion of some aspect of laboratory practice is needed. Topics covered in the various volumes are as follows:

Vol. I: Physical Methods of Organic Chemistry, 3rd Ed.; in four parts
Vol. II: Catalytic, Photochemical, and Electrolytic Reactions, 2nd Ed.
Vol. III: Separation and Purification (part I); Laboratory Engineering (part II); 2nd Ed.
Vol IV: Distillation
Vol. V: Adsorption and Chromatography
Vol. VI: Micro and Semimicro Methods
Vol. VIII: Organic Solvents, 2nd Ed.
Vol. VIII: Investigation of Rates and Mechanisms of Reactions, 2nd Ed.; in two parts
Vol. IX: Chemical Applications of Spectroscopy
Vol. X: Fundamentals of Chromatography
Vol. XI: Elucidation of Structures by Physical and Chemical Methods; in two parts

Other general reference works used very frequently in organic research are given below under headings which indicate their major uses.

To learn if a compound is commercially available, and if so, where it can be purchased:

Chem Sources. Flemington, N.J.: Directories Publishing Co. This annual covers the more than 30,000 organic compounds sold by over 600 companies. It does not give price information, however, and thus it is still desirable to have the catalogs of some of the major suppliers, e.g., Aldrich Chemical Co., Milwaukee, Wis. 53210; Alfa Inorganics Inc., Beverly, Mass.

---

† A. Weissberger, Ed., *Technique of Organic Chemistry,* New York, Interscience Publishers. A corresponding work in German is also available: E. Müller, Ed., *Methoden der Organischen Chemie (Houben-Weyl)*, Vols. 1–14, Stuttgart: Georg Thieme Verlag.

01915; Eastman Organic Chemicals, Rochester, N.Y., 14603; Fluka AG, Buchs, Switzerland; K & K Laboratories, Inc., 121 Express St., Plainview, N.Y. 11803; Matheson, Coleman, and Bell, 2909 Highland Ave., Norwood, Ohio 45212.

To check the usual physical properties of common organic compounds:

Handbook of Chemistry and Physics. Cleveland, Ohio: The Chemical Rubber Company. This annual, which changes relatively little from year to year, contains constants for about 14,000 organic compounds, as well as for many inorganic and organometallic compounds. These listings are accompanied by numerous useful mathematical, physical, and chemical tables, which combine to make this the book referred to most often by organic chemists.

Handbook of Chemistry. New York: McGraw-Hill Book Company, N.A. Lange, Ed. Essentially paralleling the handbook above, this one is revised about once every five years. Its tables of constants cover about half as many compounds as the CRC handbook.

Merck Index. Rahway, N.J.: Merck & Co., Inc. This book, revised about once every 10 years, lists physical and physiological properties for about 10,000 chemicals of interest to the pharmaceutical industry.

Dictionary of Organic Compounds. Oxford, England: Oxford University Press, 4th Ed., 1965, with annual supplements. This five-volume work, covering over 40,000 compounds, gives molecular and structural formulas, melting and boiling points, recrystallization solvents, uses, and literature references.

To find whether a compound has been reported previously, or to learn anything published regarding it:

Chemical Abstracts. Easton, Pa.: American Chemical Society. Abstracting virtually all papers and patents, this invaluable weekly publication, only a few months behind the original literature, has semiannual author, subject, formula, and patent indices. The seven collective indices cover the periods 1907–16, 1917–26, 1927–36, 1937–46, 1947–56, 1957–61, 1962–66. To aid in finding information in the latest issues before the semiannual indices have appeared, the material in each issue is grouped topically, and a keyword index is included with each issue.

Beilstein's Handbuch der Organischen Chemie. Berlin: Springer Verlag. This comprehensive work surveys all organic compounds which have been characterized, comparing methods of preparation. It suffers, however, from being many years behind the original literature. The initial volumes cover the literature to 1910, the first supplement to 1920, the second to 1930, the third to 1940, and the fourth to 1950. The organization is rather tricky, and a perusal of E. H. Huntress, *A Brief Introduction to the Use of Beilstein's Handbuch der Organischen Chemie,* New York, John Wiley & Sons, Inc., 1938 or O. A. Runquist, *A Programmed Guide to Beilstein's Handbuch,* Minneapolis, Burgess Publishing Co., 1966, is recommended before the first use of Beilstein.

To find precedents for a reaction:

Reagents for Organic Synthesis. New York: John Wiley & Sons, Inc., 1967, by L. F. Fieser and M. Fieser. This valuable book gives preparations, properties, and uses for over 1000 reagents used in organic chemistry.

Organic Syntheses. New York: John Wiley & Sons, Inc. These annual volumes contain detailed procedures, checked in a laboratory other than the submitter's, for the preparation of specific compounds. Every 10 years, the decade's preparations are revised and combined in a collective volume, the fourth of which appeared in 1963. The reactions are indexed according to the reaction type as well as by the specific compounds whose preparations are given.

Organic Reactions. New York: John Wiley & Sons, Inc. In these volumes, which appear about every other year, certain general reactions are discussed thoroughly with regard to scope and best conditions. An attempt is made to include *each literature report* of the reaction. Over 100 reactions have been covered to date.

Synthetic Organic Chemistry. New York: John Wiley & Sons, Inc., 1953, by R. B. Wagner and H. D. Zook. This volume gives sample procedures and copious references for most types of organic reactions prior to 1951.

Synthetic Methods of Organic Chemistry. Basel, Switzerland: S. Karger AG, by W. Theilheimer. These annual volumes summarize new reactions and procedures and provide literature references. A cumulative index appears with every fifth volume.

To order chemical supplies and equipment:

Laboratory Guide to Instruments, Equipment and Chemicals, published annually by the American Chemical Society, Easton, Pa. 18042, costs $2 and gives names and addresses of suppliers. It contains (among others) sections entitled "Who Makes Instruments and Equipment," "Who Makes Chemicals and/or Offers Services," "Laboratory Supply Houses," and "Company Addresses." It names companies which supply certain products, but it is no substitute for company catalogs. Catalogs can be obtained free from the major suppliers, e.g., Fisher Scientific Company, 633 Greenwich St., New York, N.Y. 10014; LaPine Scientific Company, 6001 S. Knox

Avenue, Chicago, Ill. 60629; E. H. Sargent and Company, 4647 W. Foster Ave., Chicago, Ill. 60630; and Arthur H. Thomas Company, Box 779, Philadelphia, Pa. 19105. Unless otherwise indicated, supplies and equipment mentioned in this book can be found in the catalogs of one of these supply houses.

To keep abreast of the chemical literature:

Annual Reports (London: Chemical Society), *Angewandte Chemie, International Edition in English* (monthly, New York: Academic Press), and *Chemistry & Industry* (weekly, London: Society of Chemical Industry) provide current summaries of progress in organic chemistry. *Chemical Abstracts* provides abstracts of virtually all developments relating to chemical progress and is divided into sections so that articles pertaining to a particular area of interest can easily be located. *Current Chemical Papers* (London: Chemical Society), *Chemical Titles* (Easton, Pa.: American Chemical Society), and *Current Contents* and *Index Chemicus* (both published by the Institute for Scientific Information, Philadelphia, Pa.) provide fast coverage of current research publications. *Chemical Reviews* and *Accounts of Chemical Research* (Easton, Pa.: American Chemical Society), *Quarterly Reviews* (London: Chemical Society), and *Angewandte Chemie* contain extensive review articles on certain topics of current interest.

To aid in writing journal articles:

Handbook for Authors of Papers in the Journals of the American Chemical Society, available for $2 from American Chemical Society Publications, 1155 Sixteenth St. N.W., Washington, D.C. 20036. The first edition, 1967, includes tables of standard journal abbreviations, dimension abbreviations, and proofreaders marks, plus much other useful information.

Contents

1

REACTION TECHNIQUES　1

2

ISOLATION TECHNIQUES   42

3

STRUCTURE DETERMINATION TECHNIQUES   98

1
Reaction
Techniques

In an introductory organic chemistry laboratory course, the student learns how to run organic reactions in "cookbook" fashion. The transition from this level of achievement to the competence required to carry out a synthesis from a vaguely defined procedure in a research journal or to develop a new synthesis is substantial. The general information given in Secs. 1.1 to 1.7 should help the student to bridge the gap between organic reactions at introductory and advanced levels. It is supplemented in Sec. 1.8 by a series of "case studies" in which the experimental approaches that were used to solve some special problems are examined.

1.1 LIBRARY PRELIMINARIES

When an unusual organic substance is desired for some purpose, the place to look first is *Chem Sources* (see Introduction) to see if it is commercially available. If it is not, it must be synthesized, and the next place to look is *Chemical Abstracts* to see if it has been prepared previously. If it has, the best literature preparation should be used unless the chemist thinks a new approach will be superior.

In following literature procedures for organic reactions, one soon learns that the yield obtained is usually less than that recorded, and in some cases none of the reported product is obtained. Fortunately, there is an important exception to this sad state of affairs: *Organic Syntheses* preparations, which have been carefully checked by chemists other than the submitters *using the submitters' recipe*, usually proceed as described. A main difficulty in reproducing literature yields is that the procedure often has been inadequately described; some vital detail is left out. Chemists should strive in recording their experimental results to achieve the proper balance between conciseness and the inclusion of sufficient detail to permit others to reproduce their results.

If the compound has not been described previously, the chemist must use his knowledge of organic reactions and his ingenuity to design an efficient synthesis from available starting materials.† If a multistep synthesis is

† See R. E. Ireland, *Organic Synthesis,* Englewood Cliffs, N.J., Prentice-Hall, Inc., 1969.

required, it is better to have any low yield reactions early in the sequence to avoid having to carry large amounts of material through the intermediate steps. It is strongly advisable to save a small amount of each intermediate isolated during such a synthesis for future reference.

In designing a procedure for the preparation of a new substance, it helps to know the reaction mechanism, the side reactions, the influence of catalysts, the time and temperature usually required, the solvent that should be used, and the isolation techniques that are best employed to purify the desired product. Information on these points can usually be obtained from an article on the general reaction in *Organic Reactions,* from the description of an analogous reaction in *Organic Syntheses,* or from one of the related works mentioned in the *Introduction.*

1.2 SCALE

In the synthesis of a large quantity of a compound, the experimental aspects of the problem are divided into two parts. In the first phase of the investigation, a series of *exploratory* reactions is carried out to determine the optimum reaction conditions; the second phase involves the extrapolation of these findings to a *preparative* scale.

In general, *preparative* reactions should be run on the largest scale compatible with the equipment available; secondary considerations, such as the volume of solvent required for subsequent extractions, must also be taken into account. If the starting material is particularly expensive or of limited availability, it is wise not to risk more than half of it in a single preparative reaction since the reaction flask may break or some similar disaster may make the reaction a total loss. On the other hand, there is no point in running *exploratory* reactions on a scale larger than the minimum necessary to get the desired yield data. With current micro methods, such reactions can be carried out conveniently on 100 mg, and sometimes on as little as a milligram. It is usually possible to run several concurrent exploratory reactions under slightly different conditions when the best conditions for a reaction are being sought.

1.3 REACTION VESSELS

1.3.1 Large Scale Reactions: Preparative reactions are usually conducted in three-necked round-bottomed flasks with standard taper ($\mathbf{T}$) ground-glass joints; these flasks are commonly available in sizes between 50 ml and 12 liters. In the larger flasks all three necks are vertical, but in the smaller sizes the two outer necks point slightly outward to facilitate the mounting of bulky units such as condensers and addition funnels.

The use of equipment fitted with standard taper joints simplifies the construction of an experimental setup and allows considerable flexibility with a few basic components. Figure 1-1 illustrates some of the commercially available standard taper reaction vessels and accessories. The use of standard taper ware is particularly desirable for operations conducted at reduced pressure since a greased standard taper joint is reasonably vacuum tight.

A disadvantage in using ground-glass joints is that under certain conditions they are liable to "freeze." This danger can be minimized by applying a small amount of stopcock grease to the upper part of the joint or by using a Teflon sleeve which slips between the two pieces of glass and gives a tight seal. In a vacuum system, it is absolutely essential that all joints be well greased in order to maintain a low pressure and avoid freezing the joints. The likelihood of freezing is also great in reactions involving alkali. A frozen joint can usually be freed by tapping with the back of a wooden-handled spatula, heating the joint with steam, or (as a last resort) removing any flammable solvent, heating the joint in a flame, and tapping occasionally.

For reactions that involve prolonged exposure of the glassware to alkaline conditions, many of the reaction vessels in the figure are available in special alkali-resistant glass. If prolonged heating with caustic is required, a plastic† or copper flask should be used.

Figure 1-2 depicts a typical large scale reaction setup in which provision is made for the controlled addition of one reactant from a dropping funnel to a second which is contained in the reaction flask. A Teflon paddle stirrer is used to agitate the reaction mixture and the apparatus is fitted with a condenser so that the reaction can be heated at reflux. The vacuum-nitrogen system permits the reaction to be carried out under an inert atmosphere (Sec. 1.6.1.)

1.3.2 Small Scale Reactions: When the reaction mixture occupies a few milliliters, the reaction can be carried out in a tapered centrifuge tube to facilitate recovery of a solid or liquid product after removal of solvent. On a still smaller scale, sections of small glass tubing sealed at one end are often used, and the course of the reaction may be followed by a spectral technique. For example, if an exploratory reaction can be conveniently followed by nuclear magnetic resonance (NMR) spectroscopy (Sec. 3.3.1), the reaction can be performed in an NMR sample tube and its progress checked periodically by NMR.

In small scale reactions, it is particularly important to avoid contamination with stopcock grease; while 50 mg of grease dissolved from a ground-glass joint will probably not be a serious contaminant in 100 g of product, it will almost surely be so in 50 mg of product.

---

† Many laboratory items made from polyethylene, polypropylene, Teflon, and other plastics are listed in the *Plastic Ware Catalog,* Cole-Parmer Instrument Co., 7330 N. Clark St., Chicago, Ill. 60626, and the *Nalgene Labware Catalog,* J & H Berge, Inc., 4111 S. Clinton Ave., S. Plainfield, N.J. 07080.

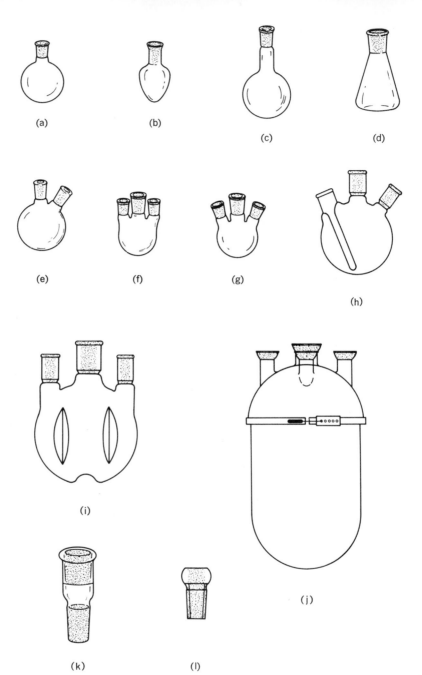

Fig. 1-1 Standard taper glassware for organic reactions: (a) round-bottomed flask; (b) pear-shaped flask; (c) long-necked, round-bottomed flask; (d) Erlenmeyer flask; (e) two-necked, round-bottomed flask; (f) three-necked, round-bottomed flask; (g) radial three-necked, round-bottomed flask; (h) two-necked, round-bottomed flask with thermometer well; (i) Morton flask; (j) resin flask; (k) enlarging adapter tube; (l) bushing-type adapter tube.

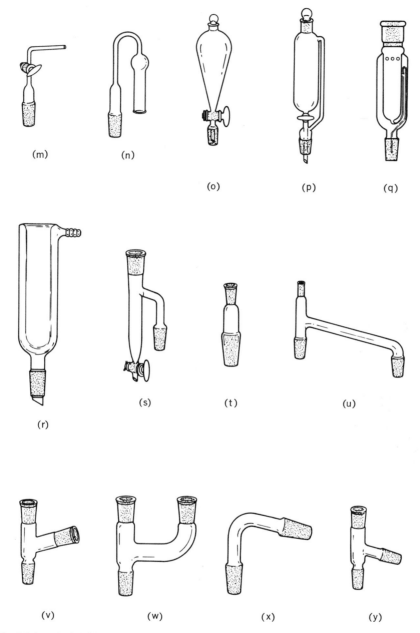

Fig. 1-1 (cont.): (m) 90° connecting tube with stopcock; (n) inverted terminal drying tube; (o) pear-shaped separatory funnel; (p) cylindrical separatory funnel with pressure equalizing tube; (q) Soxhlet extraction tube; (r) Dry Ice condenser; (s) Barrett distilling receiver (Dean-Stark trap); (t) reducing adapter tube; (u) three-way connecting tube; (v) vertical delivery distilling tube; (w) three-way connecting tube; (x) 75° connecting tube; (y) three-way connecting tube.

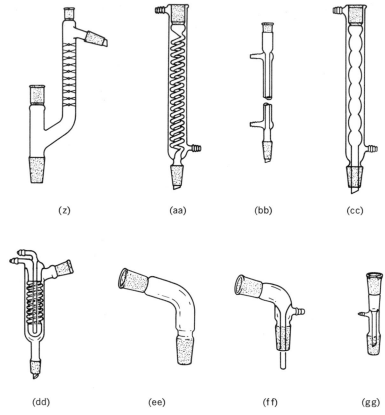

(z)   (aa)   (bb)   (cc)

(dd)   (ee)   (ff)   (gg)

Fig. 1-1 (cont.): (z) Claisen distilling head with Vigreux column; (aa) Graham condenser; (bb) West condenser; (cc) Allihn condenser; (dd) Friedrichs condenser; (ee) collecting adapter; (ff) vacuum adapter; (gg) straight vacuum adapter.

1.3.3 Pressure Reactions: Certain reactions such as some Diels-Alder reactions and catalytic hydrogenations involve working with gases at pressures greater than 1 atm. Under these circumstances the reaction must be conducted in a sealed reaction vessel. While ordinary glass reaction setups like the one in Fig. 1-2 are able to hold a nearly perfect vacuum without imploding, the equipment is not designed to withstand pressure from within and the apparatus may explode if it is sealed off and the internal pressure is allowed to rise.

For reactions in which the pressure will not exceed 20 atm, sealed heavy-walled Pyrex tubes can be used as reaction chambers. Since there is always danger of an explosion under these conditions, it is imperative that the tube be handled only with thick gloves from behind a sturdy safety shield while the tube is under pressure.

A method for sealing a tube for a high pressure reaction is illustrated in Fig. 1-3. The contents may be cooled in a Dry Ice or liquid nitrogen bath to prevent ignition during the sealing process. A glass rod (~ 12 mm) is sealed to the open end of the tube and the tube is heated about 1 in. from the open end while it is rotated. As the glass melts, it is allowed to thicken somewhat and to collapse to about one-third of the original tube diameter and then it is drawn to give a thick-walled capillary. After cooling, the capillary is sealed in the flame. If an inert atmosphere is desired in the tube, the capillary is broken before sealing, rubber tubing connected to a three-way stopcock is slipped over the open capillary end, and the tube is successively evacuated and filled with inert gas.

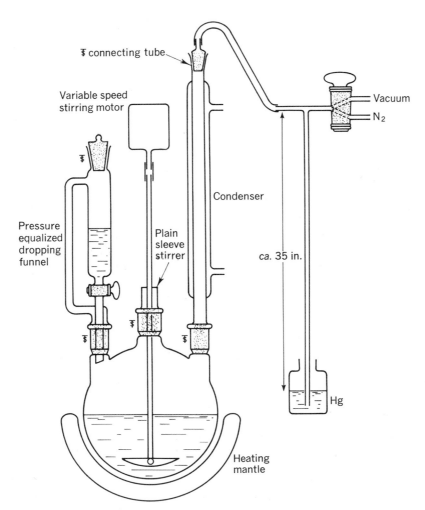

Fig. 1-2 Large scale reaction setup.

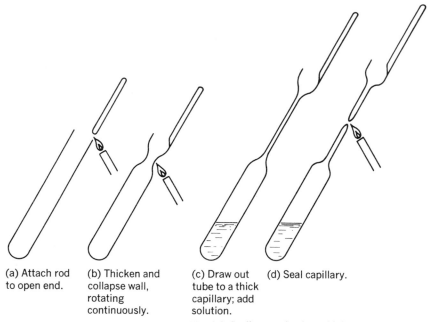

(a) Attach rod to open end.

(b) Thicken and collapse wall, rotating continuously.

(c) Draw out tube to a thick capillary; add solution.

(d) Seal capillary.

Fig. 1-3 Sealing a tube for a high pressure reaction.

To carry out the reaction, the tube should be placed in a steel jacket which is open at one end, and the jacket is then inserted in a Carius furnace so that the tube and its contents can be heated to the desired reaction temperature. If a furnace is not available, the tube can be clamped into a well-shielded oil bath and heated in this manner.

After the reaction is complete and the tube has cooled to room temperature, it is chilled in a steel jacket down to Dry Ice temperature. This should be done slowly and cautiously since thermal shock can cause a tube under pressure to explode. When the tube and its contents have been cooled, the tip of the tube is slid out of the jacket and heated in a hot flame to release any pressure within the tube. *This should only be done behind a shield and while wearing heavy asbestos gloves.*

For reactions developing pressures above 20 atm, heavy-walled steel reaction vessels ("bombs") are used. Glass liners are available for cases in which the metal would be dissolved or would adversely affect the reaction, and some of the commercial steel reaction vessels (Fig. 1-4) can be heated and rocked or stirred during the reaction. Great care must be used in working with high pressure equipment and adequate shielding is an absolute necessity. To assure maximum safety in its use, equipment of this sort should be grouped in an area well-removed from that frequented by laboratory personnel, and it is good practice to train a single individual to operate and maintain it.

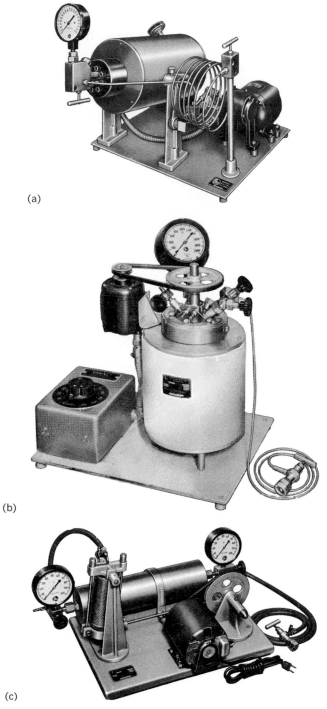

(a)

(b)

(c)

Fig. 1-4 (a) Rocked high pressure reactor; (b) stirred high pressure reactor; (c) Parr apparatus for moderate pressure reactions.

1.3.4 Pyrolyses: For reactions that require high temperatures but which take place at atmospheric pressure (e.g., acetate pyrolysis, ketene formation), it is usually convenient to use a pyrolysis apparatus (Fig. 1-5). An electrically heated cylindrical furnace (*ca.* $70) fitted with a thermocouple is clamped in a vertical position. The pyrolysis tube, which is made of Pyrex (usable up to 600°), Vycor (to 1200°), or quartz (to 1300°), is inserted; then an addition funnel, a three-necked flask, and a condenser are assembled as shown. It is usually advantageous to pack the pyrolysis tube with glass beads, glass helices, or porcelain chips to increase the surface area within the tube.

The reaction is started by bringing the pyrolysis tube, which serves as the reaction chamber, to the desired temperature and displacing the air in

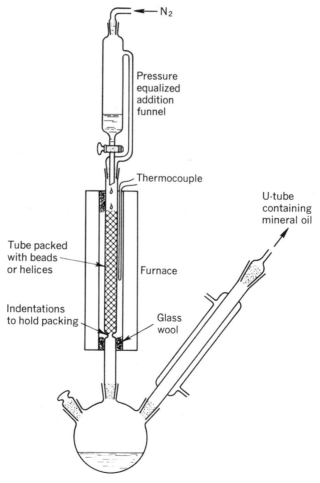

Fig. 1-5 Pyrolysis apparatus.

the system with a stream of nitrogen gas. When temperature equilibrium in the column has been reached, slow addition of the reagent is started. The products are swept down the column and condensed in the flask. A significant advantage of this type of system is that it can be operated on a continuous basis (see Sec. 1.8.6).

In some cases, the pyrolysis apparatus shown in Fig. 1-6 will suffice. For example, to convert dicyclopentadiene to cyclopentadiene,† the system in Fig. 1-6(a) is very efficient. The cold finger in the distillation head is filled with a liquid that has a boiling point higher than that of the product but well below that of the reactant. The flask containing the reactant is brought to a gentle reflux and the liquid in the cold finger soon begins to reflux. The reactant condenses on the cold finger and returns to the reaction flask, whereas the product distills and collects in the receiving flask.

For pyrolyses that require higher temperatures, a reaction flask such as that shown in Fig. 1-6(b) is useful. The reactant is placed in the flask and the air is replaced with nitrogen. The reactant is brought to a gentle reflux and the electrical coils are then heated by passing current through them. The product (if sufficiently volatile) distills from the reactor, while the unpyrolyzed reactant returns to the reaction flask. A reactor of this design has been used for the conversion of cyclohexene to 1,3-butadiene and for the pyrolysis of acetone to ketene.

1.3.5 Photochemical Reactions: In photochemical reactions, the chemist is faced with the problem of getting sufficient light of the proper wavelength into the reaction system. Since many types of glass serve as effective filters of ultraviolet radiation, the simplest general approach is to use a photo-chemical reactor with a quartz-encased bulb that can be immersed in the solution. Quartz is desirable since it is transparent to ultraviolet radiation of wavelengths above 2000 Å. Table 1-1 contains a list of optical materials and their transmission characteristics; at the thickness given in the table, 50% of the light of the indicated wavelength is absorbed and virtually all the light of shorter wavelengths. Selected regions of the ultraviolet and visible spectrum can be used by placing appropriate filters between the light source and the reactants.‡

1.3.6 Polymerizations: Polymerizations often yield insoluble products that are difficult to remove from the reaction vessel. Such a reaction is conveniently carried out in a relatively inexpensive "polymer tube" (essen-

---

† R. B. Moffett, *Organic Syntheses,* Coll. Vol. IV, p. 238.

‡ Details regarding the great variety of available lamps and filters can be obtained from General Electric Co., Schenectady, N.Y. and The Hanovia Lamp Division, Englehard Hanovia, Inc., 100 Chestnut Street, Newark, N.J. A convenient photochemical apparatus (the "Rayonet" Reactor) is available for $500 from the Southern N.E. Ultraviolet Co., Newfield St., Middletown, Conn. For an excellent discussion of experimental techniques in photochemistry, see J. G. Calvert and J. N. Pitts, Jr., *Photochemistry,* New York, John Wiley & Sons, Inc., 1966.

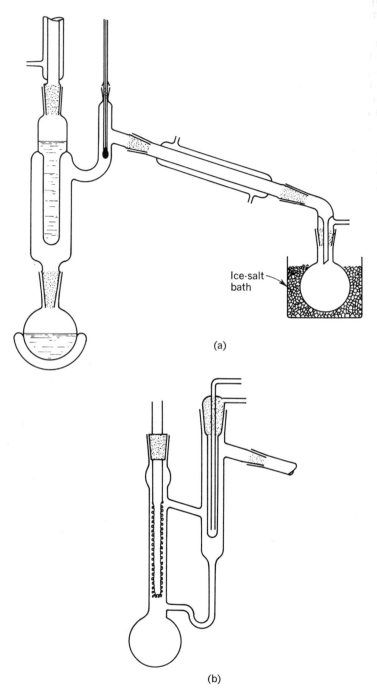

(a)

Ice-salt
bath

(b)

Fig. 1-6 Pyrolysis apparatus not requiring a furnace.

Table 1-1

OPTICAL CHARACTERISTICS OF SELECTED SUBSTANCES

Material	Thickness, mm	50% Transmission wavelength, Å
Window glass	1	3160
	3	3300
	10	3520
Pyrex (Corning 774)	1	3060
	2	3170
	4	3300
Corex D	1	2780
	2	2880
	4	3040
Corex A	2.9	2480
Vycor 791	1	2150
	2	2230
	4	2360
Quartz, clear fused (General Electric Co.)	10	1940
Suprasil (Engelhard Industries)	10	1700
Water	20	1880
	40	1920
	80	2020

tially a test tube with a constriction at the open end) that is broken to obtain the product or in a special (and expensive) "resin" flask [(Fig. 1-1(j)] that opens into two parts by virtue of a ground-glass joint around the middle.†

1.3.7 Calorimetry: Occasionally the reaction product of greatest interest is the heat given off, and it is desired to measure it accurately. A simple yet effective calorimeter has been described by Arnett and co-workers.‡

1.4 REAGENTS

1.4.1 Reagent Purity: To realize consistent and optimum results in a synthetic step, it is desirable either to use pure reagents or to know what contaminants are present. Some idea of purity can easily be ascertained in most cases by the determination of a melting point or by gas or thin layer chromatography (Secs. 2.5 and 2.6.5). Checking reagent purity is an excellent habit; contamination of reagents is probably the single most important factor contributing to reaction failure and to the wasted research time that results.

1.4.2 Addition of Liquids: The controlled addition of a liquid reagent to a reaction mixture is most easily accomplished through the use of a

† For further information on experimental methods in polymer chemistry, see W. R. Sorenson and T. W. Campbell, *Preparative Methods of Polymer Chemistry,* 2nd Ed., New York, Interscience Publishers, Inc., 1968.

‡ E. M. Arnett, W. G. Bentrude, J. J. Burke, and P. McC. Duggleby, *J. Amer. Chem. Soc.,* **87,** 1541 (1965).

dropping funnel equipped with a pressure-equalizing side arm [Fig. 1-1(p)]. Calibrated versions, useful when a carefully controlled addition is required, are also available. Dropping funnels fitted with Teflon stopcocks are preferable since they do not require grease and do not freeze. To increase the sensitivity of a Teflon stopcock for controlling the rate of addition, it is advisable to notch it slightly at both ends of the bore:

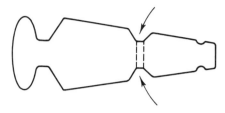

For reactions that involve high dilution conditions, a Hershberg funnel is often used (Fig. 1-14). The critical features of this funnel are a constant bore capillary glass tube and a length of inert wire of a diameter that is comparable to that of the capillary bore and is sealed into a glass rod. The solution in the funnel passes between the wire and glass tubing in a thin film and the rate of addition is controlled by the depth to which the wire is inserted in the capillary. The advantage of this type of dropping funnel is that it is ideally suited for a slow and steady dropwise addition of a solution to a reaction mixture. If extreme accuracy is desired, a motor driven syringe (Sec. 1.8.1) is preferable.

1.4.3 Addition of Solids: The controlled addition of a solid to a reaction flask presents more of a problem from an experimental point of view. The most satisfactory approach is to find a suitable solvent for the solid and carry out the addition of the solution by one of the techniques described above. An alternative which has been useful in selected instances such as hydride reductions is to place the solid in a simple (Fig. 1-7) or Soxhlet [Fig. 1-1(q)] extractor above the reaction flask and control the rate of addition by reflux of an appropriate solvent which gradually leaches the reactant into the reaction flask. This method works best for granular materials since fine powders tend to slow the rate of passage of solvent to such an extent that flooding of the extraction chamber occurs. Under these circumstances, mixing the powder with an inert substrate such as sand or diatomaceous earth minimizes the difficulty. A disadvantage of the Soxhlet extractor is that most designs provide for intermittent siphoning of substantial volumes of solvent into the reaction flask; if the reaction between the reagents is very exothermic, as in many hydride reductions, the reaction is difficult to keep under control.

When a solid to be added directly to the reaction flask must be protected from the atmosphere, the methods illustrated in Fig. 1-8 are usually satisfactory. The first apparatus is constructed by sealing a standard taper joint

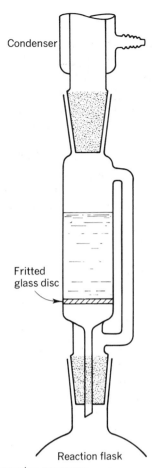

Condenser

Fritted
glass disc

Reaction flask

Fig. 1-7 Simple continuous extraction apparatus.

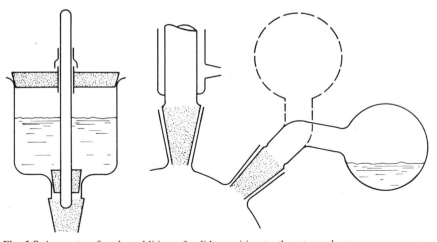

Fig. 1-8 Apparatus for the addition of solids sensitive to the atmosphere.

to the bottom of a beaker. A stopper that fits into the standard taper joint comfortably is then attached to a glass rod. A second stopper is used to seal off the top of the beaker and this is fitted with a glass sleeve that passes through the center of the stopper. A glass rod is passed through this sleeve and the sleeve and rod are connected by a flexible piece of rubber tubing. Small amounts of solid are added by raising the glass rod and gently tapping the sides of the beaker.

The second apparatus in Fig. 1-8 uses a tip flask, constructed by sealing a male joint to a flask through a gooseneck bent at about a 135° angle. By rotating the flask from a nearly horizontal position toward the vertical position, the solid can be introduced gradually into the reaction vessel. A less sophisticated variation of this approach is to place the solid in an Erlenmeyer flask and connect this to the reaction flask with a piece of large diameter (Gooch) rubber tubing. The solid is then added by tipping the Erlenmeyer flask and shaking it gently.

1.4.4 Addition of Gases: Gaseous reagents are sometimes generated *in situ* (B_2H_6, N_2H_2) or in a separate apparatus (ozone,[†] ketene[‡]), but the most convenient source of many gases is a commercial cylinder. Gases available in cylinders include the rare gases, two grades of nitrogen (regular, >99.6% pure; and prepurified, >99.996% pure), oxygen, ozone, fluorine, chlorine, hydrogen fluoride, hydrogen chloride, hydrogen bromide, hydrogen cyanide, boron trifluoride, sulfur tetrafluoride, phosgene, most of the gaseous alkanes and alkenes, cyclopropane, 1,3-butadiene, allene, acetylene, methylacetylene, ethylacetylene, a large number of halogenated alkanes and alkenes, carbon monoxide, carbon dioxide, ethylene oxide, ammonia, several gaseous amines, and methyl mercaptan. The pressure in a full cylinder of certain gases that are not readily liquified is as high as 2500 psi, and for these gases it is desirable to employ a pressure reducing valve ($20–70) rather than a simple valve ($5–15) that could allow full cylinder pressure to reach the reaction system. Gas cylinders are potentially dangerous, especially those containing highly toxic or flammable substances, and they should be strapped securely to prevent being knocked over and having the valve broken off.

Gaseous reagents are usually introduced via a glass tube below the surface of a liquid in the reaction flask. To provide extra contact between gas and liquid, the gas can be passed in as small bubbles with the aid of a tube with a fritted glass end.

1.4.5 Solvents: A primary consideration in any reaction is what solvent, if any, should be used. A solvent serves two primary functions, namely, to allow the reaction to take place under homogeneous conditions and to

---

† Ozonators are available from the Ozone Processes Division, Welsbach Corp., Philadelphia, Pa. 19129, and from Purification Sciences, Inc., 75 E. North St., Geneva, N.Y. 14456, for about $500.

‡ Conveniently prepared using the method of J. W. Williams and C. D. Hurd, *J. Org. Chem.*, **5**, 122 (1940); see also the apparatus described in Sec. 1.3.4.

control the reaction temperature. Whenever it is possible and experimentally practical, a reaction should be carried out under homogeneous conditions since the results tend to be more reproducible and the reaction will occur more rapidly. By choosing a solvent that boils at the desired reaction temperature and by running the reaction in the refluxing solvent, the system is provided with a built-in thermostat. If a reaction is exothermic and a solvent is omitted, local heating can be excessive and result in decomposition of reactants or products and the reaction may get out of control.

Solvents can also grossly affect the course of a reaction. For example, a nucleophilic substitution can occur with clean inversion of configuration in a nonpolar solvent and with complete racemization in a polar solvent. Certain polar aprotic solvents such as dimethyl sulfoxide, dimethylformamide, and sulfolane efficiently solvate cations but not anions; as a result, bases and nucleophiles exhibit increased basicity and nucleophilicity in these solvents. This property of these solvents has great importance in synthesis since it permits many base-catalyzed reactions to proceed at much lower temperatures than are normally necessary.†

Another interesting use of reaction solvents is frequently made in reactions that involve an equilibrium in which one of the products is water (e.g., ester formation, enamine synthesis, dehydration of hydrates). Under these circumstances, selection of a solvent such as benzene or carbon tetrachloride permits the rapid removal of the water on a continuous basis by azeotropic distillation (see Sec. 1.8.3).

An ideal solvent should be inexpensive, easily purified, readily separable from the reaction products, and inert to the reagents that are employed. A list of frequently used reaction solvents is given in Table 1-2. The dielectric constant is included to provide an approximate measure of solvent "polarity." Most of the purification procedures given were developed before the advent of molecular sieves,‡ and if water is the offending impurity, the best purification process probably consists of refluxing for a few hours over sieves, which have very great affinity for water. For example, Type 4A (meaning 4 Å diameter pores) sieves have been recommended for drying dimethylformamide, pyridine, and dimethyl sulfoxide, and Type 5A for tetrahydrofuran and dioxane.

1.5 AGITATION

Since efficient mixing of the reactants is often critical for the success of a reaction, provision for adequate stirring is frequently of paramount importance. For ordinary work on a moderate to large scale, a Teflon paddle on a ground-glass shaft [Fig. 1-9(a)] driven from above by an electric motor is usually satisfactory. The shafts and bushings are made with sufficient

† For example, see R. S. Kittila, *Dimethylformamide Chemical Uses,* Wilmington, Del., E. I. du Pont de Nemours & Co., 1967.
‡ Sieves and a bulletin on their use are available from Union Carbide Chemicals Co., Union Carbide Corp., 270 Park Ave., New York, N.Y. 10017.

Table 1-2

REACTION SOLVENTS

Solvent	Boiling point, °C	Melting point, °C	D_4^{20}	Dielectric constant at ~25°	Purification reference
Polar, protic					
Ammonia	−33.4	−77.7	0.682 (−33°)	22.4 (−33.4°)	—
Methanol	65	−97.8	0.787 (25°)	32.6	1
Ethanol	78.5	−117.3	0.789	24.3	2
tert-Butyl alcohol	82.3	25.5	0.786	11.5	2
Water	100.0	0.0	0.998	78.5	—
1-Butanol	117	−89.5	0.802	17.1	3
Acetic acid	118.5	16.6	1.049	6.2	4
Ethylene glycol monomethyl ether (methyl cellosolve)	125	−85.1	0.968 (15°)	—	—
Diethylene glycol monomethyl ether (methyl carbitol)	193	< −84	1.035	—	—
Ethylene glycol	198	−13.2	1.117 (15°)	37	5
Diethylene glycol	245	−10.5	1.118	—	—
Polar, aprotic					
Acetonitrile	80.0	−45.7	0.777 (25°)	38	6
Pyridine	115.5	−42	0.988 (15°)	12.3	7
Dimethylformamide (DMF)	153	−61	0.945 (25°)	36.7	8
Dimethyl sulfoxide (DMSO)	189 (~100 dec)	18.5	1.101	45	9
Quinoline	237.1	−15.9	1.096 (17°)	9.0	10
Sulfolane	285	28	1.261 (30°)	—	11
Ethers					
Diethyl ether	34.6	−116.2	0.708 (25°)	4.3	12
Tetrahydrofuran (THF)	65	−65	0.888	7.9	13
Ethylene glycol dimethyl ether (monoglyme)	83	−58	0.869 (15°)	—	—
Dioxane	101	11.8	1.034	2.2	14
Diethylene glycol dimethyl ether (diglyme)	161.5	−68	0.945	—	11
Triethylene glycol dimethyl ether (triglyme)	216	−45	0.990	—	—
Tetraethylene glycol dimethyl ether (tetraglyme)	275	−27	1.009	—	—
Alkyl halides					
Methylene chloride	41	−97	1.335 (15°)	9.1	15
Chloroform	61.2	−63.5	1.489	4.8	16
Carbon tetrachloride	76.8	−23.0	1.584 (25°)	2.2	17
Tetrachloroethylene	121.2	−22.4	1.631 (15°)	2.6	18
Chlorobenzene	131.7	−45.6	1.106	5.6	17
Hydrocarbons					
Pentane	36	−129.7	0.626	1.8	19
Benzene	80.1	5.5	0.874 (25°)	2.3	20
Cyclohexane	81	6.5	0.779	2.0	21
Toluene	110.6	−95	0.867	2.4	20
p-Xylene	138	14	0.854 (28°)	2.3	20
Decalin (mixture of stereoisomers)	185–195	−124	0.88	2.0	22
Biphenyl	255.3	69.2	1.041	2.5	23

References for Table 1-2

1. E. C. Evers and A. G. Knox, *J. Amer. Chem. Soc.* **73,** 1739 (1951).

2. A. A. Maryott, *ibid.* **63,** 3079 (1941).

3. C. P. Smyth and W. N. Stoops, *ibid.* **51,** 3312, 3330 (1929).

4. A. W. Hutchinson and G. C. Chandlee, *ibid.* **53,** 2881 (1931).

5. C. P. Smyth and W. S. Walls, *ibid.* 2115.

6. G. L. Lewis and C. P. Smyth, *J. Chem. Phys.* **7,** 1085 (1939).

7. D. Jerchel and E. Bauer, *Angew. Chem.* **68,** 61 (1956).

8. H. J. Ferrari and J. G. Heider, *Microchem. J.* **7,** 194 (1963).

9. H. O. House and V. Kramar, *J. Org. Chem.* **28,** 3376 (1963).

10. O. Kruber, *Chem. Zentr.* 1378 (1955).

11. L. F. and M. Fieser, *Reagents for Organic Synthesis,* New York, John Wiley & Sons, Inc. 1967.

12. A. Weissberger and E. S. Proskauer, *Technique of Organic Chemistry,* Vol. VII, New York, Interscience Publishers, 1955, p. 367.

13. M. Pestemer, *Angew. Chem.* **63,** 122 (1951).

14. C. A. Kraus and R. M. Fuoss, *J. Amer. Chem. Soc.* **55,** 21 (1933).

15. J. H. Mathews, *ibid.* **48,** 562 (1926).

16. P. M. Gross and J. H. Saylor, *ibid.* **53,** 1744 (1931).

17. D. R. Stull, *ibid.* **59,** 2726 (1937).

18. J. Timmermans and Lady Hennaut-Roland, *Chem. Zentr.* 236 (1931).

19. H. Ley and H. Hunecke, *Ber.* **59,** 523 (1926).

20. B. J. Mair, D. J. Termini, C. B. Willingham, and F. D. Rossini, *J. Res. Nat. Bur. Std. A* **37,** 229 (1946).

21. R. W. Crowe and C. P. Smyth, *J. Amer. Chem. Soc.* **73,** 5406 (1951).

22. R. Ziegler, *Z. Phys.* **116,** 716 (1940).

23. A. J. Streiff, L. H. Schultz, A. R. Hulme, J. A. Tucker, N. C. Krouskop, and F. D. Rossini, *Anal. Chem.* **29,** 361 (1957).

precision that they can be used under reduced pressure. To assure a long lifetime for the stirrer, care should be taken to align the stirring motor, shaft, and bushing, and the shaft must be lubricated before each use and intermittently throughout prolonged operation. Another type of stirrer often used for this scale reaction is a Hershberg stirrer [Fig. 1-9(b)], made from a glass rod and Nichrome wire.†

A variety of motors are available that can be used to vary the rate of

---

† For a more elaborate design for a Hershberg stirrer, see P. S. Pinkney, *Organic Syntheses,* Coll. Vol. II, p. 117.

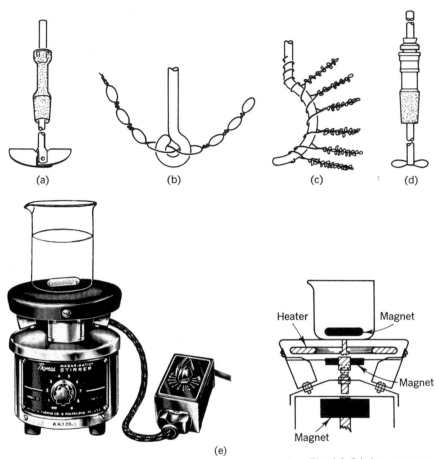

(a)   (b)   (c)   (d)

(e)

Fig. 1-9 Stirring apparatus.

stirring over a wide range. The rate of rotation can be controlled via a rheostat in series with the motor. When the reaction involves stirring a solid dispersed in a liquid, it is important to use a powerful stirrer to avoid burning out the motor. If the sparking that occurs in an ordinary electric motor presents a serious fire hazard, an air or water driven motor or a "sparkless" electric motor can be used instead.

A major area of flexibility for the type of stirring arrangement above lies in the design of the propeller. If the reaction involves copious amounts of a solid, the simple Teflon paddle or Hershberg stirrer is usually not adequate to bring about a uniform dispersal of the solid. During reflux, this results in bumping (which can actually break the glass apparatus) and possible decomposition of the reaction products due to local overheating. A stirrer that has been found to be particularly useful in these circumstances is shown in Fig. 1-9(c).† A stiff wire is wrapped around the stirring shaft

† H. R. Snyder, L. A. Brooks, and S. H. Shapiro, *Organic Syntheses,* Coll. Vol. II, p. 531.

as shown and at frequent intervals "tails" long enough to touch the sides of the flask are twisted off. During operation, these tails continuously scrape the sides of the flask and provide maximum agitation. Another device that may work when a paddle stirrer is insufficient consists of a metal propeller on a metal shaft [Fig. 1-9(d)], driven by a high-speed motor.

For reactions that can be conducted at room temperature and require vigorous agitation, a Waring Blendor, available in 1 quart and 1 gallon sizes, makes a useful reaction vessel. It provides extremely efficient stirring, particularly if solids are present; and it has the further advantage of being provided with a lid to prevent spillage. An ordinary Waring Blendor should not be used with highly flammable solvents such as ether or pentane due to the fire hazard.

On a smaller scale, magnetic stirrers with Teflon-coated stirring bars are widely used. Commercial magnetic stirrer-hot plate and/or cool plate combinations [Fig. 1-9(e)] are useful for reactions that can be run in beakers or Erlenmeyer flasks. Some caution must be exercised at high stirring speeds since the stirring bar can get out of phase and be thrown through the side of the flask.

For small scale reactions that require vigorous stirring or uniform dispersal of a solid, a vibrating stirrer is most efficient. This consists of a shaft which has a fluted disc welded to its end and a motor that vibrates the shaft through a very small period at high frequency. This establishes a continuous shock wave through the solution and results in vigorous agitation of the reactants.†

The degree of effective agitation can also be influenced by the flask design. A Morton flask [Fig. 1-1(i)] is a round-bottomed flask with several large creases strategically placed to serve as baffles for the solution and increase the agitation by several orders of magnitude.

1.6 ATMOSPHERE OVER REACTION

1.6.1 Inert Atmospheres: The oxygen, moisture, and carbon dioxide in air very often lower yields considerably and for this reason many reactions are run under an inert atmosphere. The most commonly used inert atmosphere is nitrogen, although argon (more dense and less reactive) and other gases are sometimes preferable. In a typical system (Fig. 1-2), nitrogen from a cylinder is passed through a three-way stopcock into the reaction vessel (often entering at the top of a reflux condenser). The system is flushed thoroughly with nitrogen at the start; to do this efficiently, the system is alternately evacuated and filled with nitrogen several times by manipulations of the three-way stopcock. The nitrogen flow is then reduced to the point where it is bubbling very slowly through the mercury, so that there is a slight positive nitrogen pressure on the system. A simpler system for main-

† A stirrer of this type, sold under the name "Vibromischer," is available from Chemap AG., Männedorf, Switzerland.

taining a positive nitrogen pressure lacks the vacuum source and the 35-in. tube, and it has mineral oil rather than mercury in the trap; it suffers from the disadvantage that the system cannot be evacuated initially.

If manual operations are to be performed in an inert atmosphere, a "dry box" with arm-length rubber gloves is generally used. It is flushed with the inert gas (usually nitrogen or argon) and kept under a positive pressure of the inert gas while in use. A good rigid dry box (Fig. 1-10) can be purchased for $1000, or a plastic "glove cabinet," which has the definite disadvantage that it can not be evacuated, for $100. A polyethylene bag filled with inert gas will often suffice and is an inexpensive alternative.

1.6.2 Hydrogenations: The most frequently used reaction requiring a special atmosphere is a catalytic hydrogenation.† Catalysts are available in numerous forms and include the metal or its oxide alone or on a support such as carbon, silica, or alumina. The activity of a particular catalyst is influenced by its physical state, the presence of chemical impurities, and the type of functional group being reduced. Deposition of the catalyst on an inert surface is frequently desirable since the activity can be modified and problems associated with coagulation of the catalyst during the reaction are avoided. The choice of catalyst for a hydrogenation is the primary factor determining the degree of success of the reaction. Table 1-3 gives some of the more common hydrogenation reactions and the catalysts that can be used to effect them.

The choice of apparatus for a hydrogenation is determined by the sample size and the pressure and temperature requirements of the reaction. A convenient and readily assembled unit useful for the low pressure hydro-

† See K. L. Rinehart, *Oxidation and Reduction of Organic Compounds,* Englewood Cliffs, N.J., Prentice-Hall, Inc., 1972.

Fig. 1-10 Dry box.

Table 1-3

CAPABILITIES OF HYDROGENATION CATALYSTS

Reaction	Pt	Pd	Ni	CuCr$_2$O$_4$	Rh	Ru
C=C $\longrightarrow$ CH—CH	+	+	+		+	+
C≡C $\longrightarrow$ CH=CH		+†	+			
(C$_6$H$_5$)—R $\longrightarrow$ (C$_6$H$_{11}$)—R	+		+		+	+
C=O $\longrightarrow$ CH—OH	+		+	+	+	+
C=C—C=O $\longrightarrow$ C=C—CH—OH						+‡
C=C—C=O $\longrightarrow$ CH—CH—C=O	+	+	+			
Ar—C—O $\longrightarrow$ Ar—CH + HO		+				
(epoxide) C—C $\longrightarrow$ C—CH (OH)		+				
COOH $\longrightarrow$ CH$_2$OH						+
COOR $\longrightarrow$ CH$_2$OH + ROH	+			+		
C≡N $\longrightarrow$ CH$_2$—NH$_2$			+			
(pyridine)—R $\longrightarrow$ (piperidine)—R			+			+
R$_3$NO $\longrightarrow$ R$_3$N		+				
NO$_2$ $\longrightarrow$ NH$_2$ and other products	+	+	+		+	+
COCl $\longrightarrow$ CHO + HCl		+†				

† Palladium partially deactivated to prevent overreduction.
‡ Not thoroughly investigated for generality.

genation of milligram to gram quantities is shown in Fig. 1-11. In addition to being useful for normal small scale synthetic applications, this apparatus is suitable for the determination of hydrogenation equivalents and studies of the kinetics of hydrogenation.†

When the apparatus in Fig. 1-11 is used, the catalyst, solvent, and a Teflon-covered stirring bar are placed in the special Erlenmeyer flask. After making sure that the gas burettes are filled with liquid (usually water) from the leveling bulbs and with stopcocks B, C, and E closed and D open, the system is alternately evacuated and flushed with hydrogen (*ca.* three times) by manipulating stopcock A. The appropriate burette is then filled with hydrogen, stopcock A is closed, and the catalyst is reduced (if necessary) and saturated with hydrogen by stirring for a few minutes. A known volume of the solution to be reduced is added through the side arm by opening stopcock E, taking care not to admit any air. The initial volume of hydrogen is recorded after the internal pressure has been adjusted to atmospheric by

† Commercial setups for the same purposes are available, e.g., the Delmar-Brown units from the Coleman Instruments Division, Perkin-Elmer Corp., 42 Madison St., Maywood, Ill. 60153.

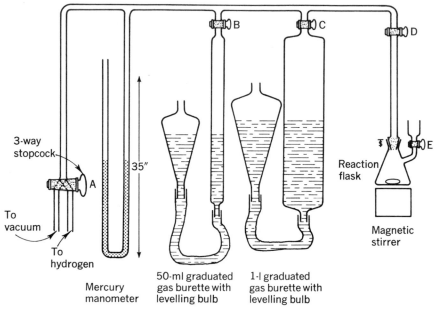

Fig. 1-11 Atmospheric pressure hydrogenation apparatus.

means of the leveling bulb, and the heterogeneous reaction is started by turning on the stirring motor. The burette is read at intervals after internal pressure has been adjusted to atmospheric with the leveling bulb. The number of moles of hydrogen taken up is calculated from the volume change, using the ideal gas law. The gas burettes can easily be refilled if necessary by opening stopcock A to the hydrogen tank and lowering the appropriate leveling bulb.

For low to medium pressure hydrogenations on a larger scale, a Parr hydrogenation apparatus [Fig. 1-4(c)] is generally used. It consists of a ballast tank which can be connected to a thick-walled bottle contained in a shaker arm; provisions for heating the reaction mixture can be made. The unit is practical for hydrogenations requiring up to 60 psi and is capable of handling up to 300 ml of solution. Since the volume is constant, the pressure gauge can be calibrated in terms of pressure drop per 0.1 mole and this offers a convenient tool for monitoring the course of the reaction.

When higher pressures or larger volumes are required, special equipment is used. The reaction vessels are usually the thick-walled steel bombs mentioned above which have provisions for heating and shaking or stirring [Fig. 1-4(a) and (b)]. It is important to keep the bombs used for hydrogenation separate from those used for other reactions since traces of certain compounds, especially those containing sulfur, will poison hydrogenation catalysts. The principle of operation of these units is identical to that just described, with the volume held constant and the reaction progress followed by the pressure drop, but greater caution must be exercised due to the high pressure involved. The literature available with each apparatus should be thoroughly read and understood before the equipment is used and, as in all hydrogenation reactions, adequate safety precautions, particularly with regard to shielding, must be observed.

In many reactions, noxious gases are produced and provision for eliminating these must be made. A simple gas absorption trap effective for the absorption of hydrogen chloride and other water-soluble gases is shown in Fig. 1-12.†

† C. S. Marvel and W. M. Sperry, *Organic Syntheses,* Coll. Vol. I, p. 95.

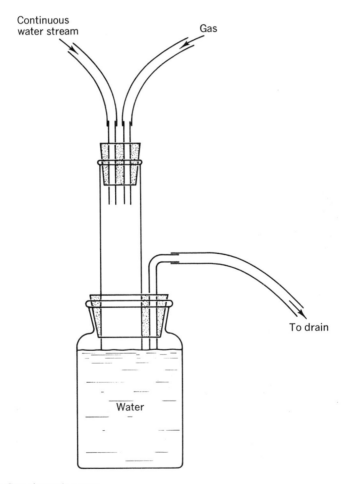

Fig. 1-12 Gas absorption trap.

1.7 TEMPERATURE CONTROL

The temperature at which organic reactions are commonly run varies from -78 to $+500°$. Accurate temperature control is very important in some of the cases in which the temperature coefficients of the desired reaction and competing side reactions differ appreciably or when the product reacts further under the reaction conditions. If heat transfer between the reaction mixture and the surroundings is efficient and the reaction does not proceed too rapidly, the reaction temperature will be close to that of the surroundings. For strongly exothermic or endothermic reactions, this desirable situation can be approached by adding one of the reagents slowly with vigorous stirring of the reaction mixture.

1.7.1 Heating: Heating is generally accomplished with an oil bath, a Wood's metal bath, or a heating mantle, all of which are electrically heated through a variable voltage transformer. An oil bath is readily assembled from a porcelain casserole, a length of high resistance wire, and mineral oil or something comparable. A commercial shielded heating coil† is shock-proof and safer to use; with this a stainless steel pan may be used in place of the casserole. In either case, the oil bath may be stirred from below using the same magnetic stirrer which rotates the stirring bar in the reaction mixture. UCON oil is the most satisfactory bath filler from room temperature to 200° since its water solubility facilitates cleaning of the equipment. In the 200–300° range, silicone oil (considerably more expensive) may be used; above this temperature, a Wood's metal bath is most convenient.

With heating mantles or the rigid "Thermowells," superheating can be a problem, and a thermocouple rather than a thermometer must be used to measure the thermostat temperature. In addition, a separate mantle ($10–40 each) is required for each size flask (sand can be used to fill in the space between the walls of the mantle and an oddly shaped flask), but they are less messy and can be operated at much higher temperatures than oil baths.

An alternative approach to heating which is excellent for reactions being carried out on a small to moderate scale is the boiling vapor bath. The apparatus is similar to the pyrolysis setup in Fig. 1-6(a) except that the reactant is placed in the inner tube and the temperature is controlled by refluxing the solvent in the outer flask. The temperature is fixed by choosing a liquid with an appropriate boiling point. Some materials which have been used for this purpose are listed in Table 1-4.

1.7.2 Cooling: For reactions run below room temperature, cooling is most easily accomplished with ice or the ice-salt mixtures listed in Table 1-5, or if lower temperatures are required, by Dry Ice-isopropyl alcohol ($-78°$) or liquid nitrogen ($-196°$). For condensing vapors below tap water

---

† $7 from Waage Electric Inc., Kenilworth, N.J. 07033.

temperature, "Dry Ice condensers" [Fig. 1-1(r)] are often used. To make most effective use of the potential of eutectic mixtures for low temperature control, it is essential that the ice be finely divided, the slush well agitated, and the bath insulated to minimize extraneous heat losses. When the eutectic baths are assembled, it is helpful to chill the salt before mixing it with the ice.

For operations which require extensive cooling periods, it is advantageous to acquire an automatic cooling unit to chill the coolant since the volume of ice that must be transported for this purpose is surprisingly large.

Table 1-4

FLUIDS FOR BOILING VAPOR BATHS

Liquid	Approximate bath temperature at reflux, °C
Acetone	55
Methanol	65
Carbon tetrachloride	75
Tetrachloroethylene	85
Water	100
Toluene	110
Chlorobenzene	130
Xylenes	140
Bromobenzene	155
o-Dichlorobenzene	180
Ethylene glycol	200
Nitrobenzene	210
Methyl salicylate	220
Biphenyl	250
Benzophenone	300
Sulfur	440

Table 1-5

EUTECTIC MIXTURES FOR ICE-SALT BATHS

Salt	Grams salt per 100 g of ice	Equilibrium temperature, °C
Potassium nitrate	12	− 3.0
Potassium chloride	24.5	− 10.7
Ammonium chloride	23	− 15.8
Sodium chloride	30.7	− 21.2
Sodium bromide	67	− 28
Magnesium chloride · 6H$_2$O	84	− 33.6
Potassium carbonate	65	− 36.5
Calcium chloride	72.5	− 54.9

A convenient and inexpensive unit for continuous operation can easily be made by stripping the compressor and cooling coils from an old refrigerator and immersing the cooling coils in the coolant solution.

1.7.3 Kinetics: Since the rate of a reaction can easily double with a 5° temperature rise, precise temperature control is a primary requirement for any kinetic study. Temperature control is usually accomplished by running the reaction with the reaction vessel immersed in a vigorously stirred and well-insulated bath (Fig. 1-13). Constant temperature is maintained in the bath through a heating or cooling element activated by an electronic relay and a thermoregulator. Commercial units are available ($250 up),† but excellent systems such as the one in Fig. 1-13 can be assembled at a fraction of their cost from components.

A bath vessel for use with noncorrosive liquids can be made by cutting the top out of a 5-liter solvent can and peening under any rough edges;

† Haake constant temperature circulators can be purchased from the Cole-Parmer Instrument & Equipment Co., 7330 N. Clark St., Chicago, Ill. 60626.

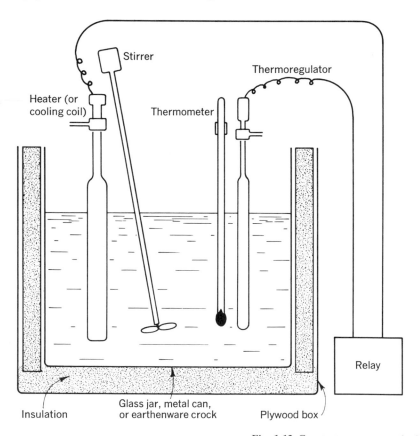

Fig. 1-13 Constant temperature bath.

Table 1-6

FLUIDS FOR CONSTANT TEMPERATURE BATHS

Bath fluid	Approximate maximum useful temperature, °C
Water	60
Motor oil	150
Tritolyl phosphate	200
Dowtherms	200
Silicone oils	150–350
Tetracresoxysilane	350–400

for corrosive bath liquids, a cylindrical glass jar is used. For insulating the bath but maintaining portability and compactness, the bath is usually placed in a box large enough to allow about 4 in. of insulating material between the bath vessel and the sides and bottom of the box. Such a box is readily constructed from $\frac{3}{8}$-in. plywood. Several aluminum support rods should be bolted to the inside of two adjacent walls of the box to clamp on accessories. The box is then filled to a depth of about 4 in. with a suitable insulating packing such as vermiculite, the bath vessel is centered in the box, and the remainder of the box is filled with insulation.

The fluid used as a heat transfer agent in the bath should have a low viscosity at the temperature employed and a high heat capacity, and preferably it should be odorless, nonvolatile, nonflammable, nontoxic, and stable to decomposition over long periods of time. A list of commercially available materials that have been used successfully appears in Table 1-6.

The most critical component of a constant temperature bath is the thermoregulator, which is usually a mercury contact thermometer. This consists of a large volume of mercury in a bulb with a capillary outlet in which a wire is inserted. The mercury in the bulb and the wire are both connected to an electronic relay; as the solution heats up, contact is made between the mercury and the wire, and the circuit to the electronic relay is completed. The relay will then turn off the heating element or turn on the cooling element, depending upon the temperature range in which the bath is being used. The temperature variation of the bath is largely determined by the quality of the thermoregulator used if the bath is well insulated and efficiently stirred.

The stirring motor must be a heavy-duty motor capable of vigorous and prolonged agitation to provide rapid heat transfer between the bath fluid and the heating element. Some experimentation is necessary to determine the best location for the stirrer in the bath to avoid the formation of a whirlpool.

Convenient heating elements are high wattage light bulbs or knife-type heaters ($10).

The operation of a constant temperature bath below room temperature requires slightly more elaborate apparatus. The previously described system can be used conveniently to about $-20°$ by running several feet of copper coil, connected to a reservoir of coolant and a circulating pump, into the bath fluid. As the bath warms up, the thermoregulator and relay activate the circulating pump and the coolant (e.g., cold brine) is circulated until the bath temperature falls just below the prescribed limit.

1.8 CASE STUDIES ILLUSTRATING SPECIAL TECHNIQUES

1.8.1 High Dilution: Although the preparation of five- and six-membered rings from appropriate acyclic precursors usually proceeds without difficulty, the synthesis of larger rings requires more elaborate techniques since bimolecular condensations tend to become more favorable than simple cyclization. To favor the unimolecular reaction at the expense of the bimolecular condensation, high dilution techniques are used, and the apparatus (Fig. 1-14) used for the Dieckmann cyclization of diethyl tetradecan-1,14-dicarboxylate is excellent for this purpose.†

$$\text{EtO}_2\text{C}(\text{CH}_2)_{14}\text{COOEt} + \text{KOt-Bu} \longrightarrow \begin{array}{c} \text{KO} \quad \text{COOEt} \\ \diagdown \quad / \\ \text{C}=\text{C} \\ / \quad \diagdown \\ \diagdown(\text{CH}_2)_{13} / \end{array} + \text{EtOH} + \text{t-BuOH}$$

After all the components have been thoroughly dried in an oven, the apparatus is assembled as shown. Dried xylene is added to the Morton flask and heated to reflux; any traces of residual moisture are removed by distilling over several milliliters and removing any azeotrope that forms in the Dean-Stark trap. Anhydrous *tert*-butyl alcohol is added to the flask, followed by potassium metal. The solution is stirred, and after all the potassium has reacted and the solution is boiling vigorously, slow addition of diethyl tetradecan-1,14-dicarboxylate from the Hershberg funnel is started. The solvent vapors, which condense and fall into the dilution chamber, dilute the diester and carry it into the Morton flask where the reaction occurs. An ethanol-xylene mixture is removed via the Dean-Stark trap at approximately the same rate as the solution is added from the dropping funnel. A slow continuous addition of the ester maximizes the probability of cyclization; a typical reaction time is 24 hr.

The reaction of sulfur dichloride with dienes can produce cyclic or polymeric products, depending upon experimental conditions.‡ To prepare

---

† N. J. Leonard and C. W. Schimelpfenig, Jr., *J. Org. Chem.*, **23**, 1708 (1958).
‡ E. J. Corey and E. Block, *J. Org. Chem.*, **31**, 1663 (1966).

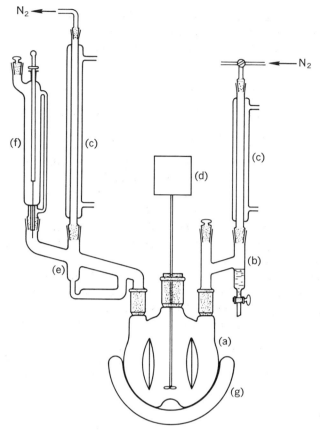

Fig. 1-14 High dilution apparatus: (a) Morton flask;   (b) Dean-Stark trap;
(c) condenser; (d) high speed stirrer; (e) dilution chamber; (f) Hershberg
addition funnel; (g) heating mantle.

2,5-bis-*endo*-dichloro-7-thiabicyclo[2.2.1]heptane from sulfur dichloride and
1,4-cyclohexadiene in good yield, it is necessary to use high dilution conditions in such a way that *neither* reagent is present in excess. This has been
accomplished using a large three-necked flask, containing methylene chloride
as the reaction solvent, fitted with a reflux condenser and two motor driven
syringes (available from JKM Instrument Co., Durham, Pa.). The syringes
are each filled with a dilute methylene chloride solution of one of the
reagents, the solvent is heated to reflux, and the programmed simultaneous
addition of the two reactants is begun. After several days of addition using
these conditions, excellent yields of the desired product are obtained. Without the use of this technique, polymeric materials are formed almost exclusively. The use of automatically driven syringes is highly recommended
whenever carefully controlled rates of addition are required.

1.8.2 Vacuum Line Reactions: Phenyldiazene (ϕN=NH) is rapidly
decomposed by traces of oxygen and, in the pure state, decomposes to

benzene and nitrogen; it can, however, be kept for long periods in solution in acetonitrile. Its reduction to phenylhydrazine by the even more unstable substance, diimide, a reaction used to characterize phenyldiazene, illustrates vacuum line techniques:[†]

$$\phi N{=}NCO_2K + H^{\oplus} \longrightarrow \phi N{=}NH + CO_2 + K^{\oplus}$$

$$KO_2CN{=}NCO_2K + 2H^{\oplus} \longrightarrow HN{=}NH + 2CO_2 + 2K^{\oplus}$$

$$\phi N{=}NH + HN{=}NH \longrightarrow \phi NHNH_2 + N_2$$

The reactions are carried out in glass apparatus (Fig. 1-15); the system is conveniently checked for leaks with a Tesla coil vacuum tester ($15). Oxygen is removed from all solutions used in the reaction by a standard "degassing" procedure, consisting of freezing and thawing the solution several times with the aid of a Dewar flask filled with liquid nitrogen while maintaining an operating pressure below 2×10^{-5} mm with a mercury diffusion pump ($60–100).

Potassium phenyldiazenecarboxylate, an acidic ion-exchange resin, lithium chloride, and a Teflon-covered magnetic stirring bar are placed in flask B via the stopper and the system is evacuated for 15 min. Dry, degassed acetonitrile is distilled from flask A into flask B simply by chilling flask B

† P. C. Huang and E. M. Kosower, *J. Am. Chem. Soc.,* **90**, 2362 (1968). Complete vacuum systems and parts for building them can be obtained from companies such as Delmar Scientific Laboratories, Inc., 317 Madison St., Maywood, Ill. and Kontes Glass Co., Vineland, N.J. 08360.

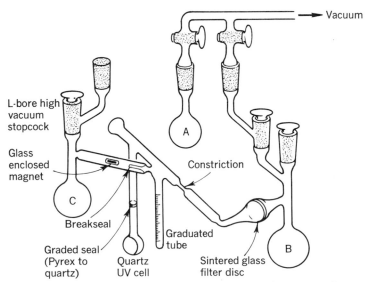

Fig. 1-15 Apparatus illustrating vacuum line techniques.

with a liquid nitrogen trap and warming flask A, and the mixture in flask B is stirred for 15 min at room temperature to form phenyldiazene. The apparatus is disconnected from the vacuum line and tilted to filter the desired volume of solution into the graduated tube. The constriction is closed with a torch and the apparatus is severed at the constriction.

Potassium azodicarboxylate and triethylamine hydrochloride are placed in flask C, and this section of the apparatus is connected to the vacuum line and evacuated. The breakseal is then broken with the magnet, and the evacuated apparatus is removed from the vacuum line and tilted to pour the phenyldiazene solution onto the diimide precursors. Phenylhydrazine formation is detected by measuring the ultraviolet spectrum of the solution via the quartz cell. Over the course of a few hours, the spectrum changes from that of phenyldiazene to that of phenylhydrazine.

1.8.3 Reactions Aided by Azeotropic Distillation: In many organic reactions, the starting materials are nearly as stable thermodynamically as the products; to obtain a good yield, it is necessary to force the equilibrium to the right by removing one of the products (not necessarily the desired one). This technique has already been illustrated in Sec. 1.8.1 by a Dieckmann condensation, forced to completion by removal of ethanol via the ethanol-xylene azeotrope.

The commonest reaction in which an azeotropic technique is useful involves the acid-catalyzed conversion of an acid to an ester. For such reactions, a simple Dean-Stark trap [Figs. 1-1(s) and 1-14] will usually suffice for removal of the water. More difficult esterifications have been successfully effected using more sophisticated equipment. For example, in the preparation of methyl pyruvate,† the position of the

$$CH_3COCO_2H + CH_3OH \overset{H\oplus}{\rightleftharpoons} CH_3COCO_2CH_3 + H_2O$$

equilibrium is particularly unfavorable for ester formation. A "methyl ester column" designed to cope with esterifications that fall into this category is illustrated in Fig. 1-16. Pyruvic acid, methanol, benzene, and a small amount of p-toluenesulfonic acid are placed in the reaction flask and heated to a vigorous reflux. The column soon fills with condensate and the azeotrope which is formed condenses and separates in the bubble chambers. The aqueous phase collects in the side arm containing the stopcock, while the benzene and methanol are continuously returned to the reaction flask. The water is drawn off as frequently as necessary. After completion of the reaction, the ester is isolated by fractional distillation.

Another situation involving azeotropic distillation, in which the *lower* layer must be returned to the reaction flask, is illustrated by the following example.

---

† A. Weissberger and C. J. Kibler, *Organic Syntheses,* Coll. Vol. III, p. 610.

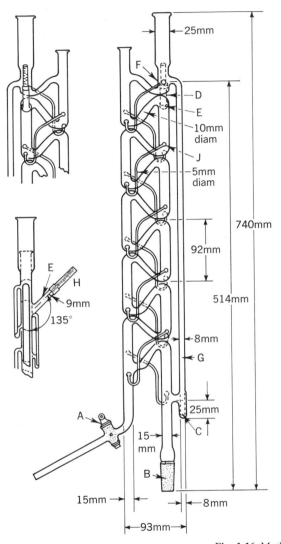

Fig. 1-16 Methyl ester column.

An efficient preparation of catechol from guaiacol involves refluxing the ether with hydrobromic acid for an extended period:†

$$
\text{(OCH}_3\text{, OH benzene ring)} + \text{HBr} \longrightarrow \text{(OH, OH benzene ring)} + \text{CH}_3\text{Br}
$$

In order to have the reaction proceed at a useful rate, concentrated hydrobromic acid (48%) is required. As the reaction proceeds, the concentration

† H. T. Clarke and E. R. Taylor, *Organic Syntheses,* Coll. Vol. I, p. 150.

of hydrogen bromide decreases and unless water is simultaneously removed, the reaction will effectively cease. Water can be removed from the reaction flask by distillation, but since guaiacol is sufficiently volatile to steam distill under these conditions, some provision must be made for separating the distilled water and guaiacol and returning the latter to the reaction flask. A continuous water separator is useful for situations such as this, and an apparatus designed for the preparation of guaiacol is shown in Fig. 1-17. Since methyl bromide vapors are highly toxic, provision is made for trapping this product.

Guaiacol and 48% hydrobromic acid are placed in a three-necked flask and the flask is fitted with a stirrer, a Vigreux column, and an automatic water separator. The solution is heated to reflux and as guaiacol, water, and methyl bromide steam distill, the first two are condensed, and collect and separate in the separation chamber. Guaiacol, which is denser than water, forms the lower layer and is forced back into the reaction flask when the hydrostatic pressure is large enough. The lighter water layer continuously

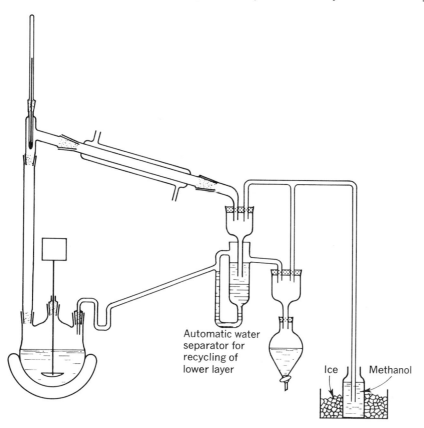

Automatic water separator for recycling of lower layer

Ice   Methanol

Fig. 1-17 Recycling apparatus for continuous removal of water.

drains into a separatory funnel and is removed occasionally. Methyl bromide evaporates continuously and is absorbed into chilled methanol contained in a bottle. The yield of purified catechol obtained by this technique is 85–87%.

1.8.4 Recycling Pyrolyses: Several simple techniques and experimental setups for pyrolyzing organic compounds have been described earlier, but, unfortunately, there are cases (especially pyrolyses of primary acetates) in which one pass through such an apparatus gives a very low conversion and this difficulty cannot be surmounted simply by raising the temperature since charring then becomes serious. If the product(s) boils sufficiently below the reactant, the recycling apparatus shown in Fig. 1-18 may solve the problem. Its use is illustrated with 2-acetoxycyclohexanone.†

The boiling flask is charged with 2-acetoxycyclohexanone and heated to reflux. The pyrolysis column, which is packed with glass beads and electrically heated to 500° with the Nichrome wire, cracks a portion of the ester; then the cyclohexenone, acetic acid, and unreacted acetate pass into the fractionating column. The products are collected by distillation and the unreacted acetate is continuously returned to the boiling flask through the pyrolysis column bypass. The vapor trap, which is merely a U-shaped portion of the bypass for the pyrolysis column, soon fills with liquid and prevents distillation around the pyrolysis chamber. The yield of 2-cyclohexenone with this apparatus exceeds 90%.

1.8.5 Soxhlet Reactions: In some syntheses, the desired product reacts with a starting material. This undesired reaction can be avoided either by separating the product from the starting material during the reaction or by forcing the product to react further, i.e., by making it a short-lived interme-diate. Both of these techniques have been used to avoid coupling products (e.g., 1,2-diphenylethane from benzyl bromide) in the preparation of allyl and benzyl Grignard reagents, and these serve as examples below.

If there is no need to isolate the Grignard reagent, the following ap-proach‡ is worth considering. For example, to prepare dimethylallylcarbinol, the magnesium turnings are covered with anhydrous ether and a small quantity of allyl bromide is added to initiate formation of the Grignard reagent.

$$CH_2{=}CHCH_2Br + Mg \longrightarrow CH_2{=}CHCH_2MgBr$$

$$CH_2{=}CHCH_2MgBr + (CH_3)_2CO \longrightarrow CH_2{=}CHCH_2C(CH_3)_2OMgBr$$

† K. L. Williamson, R. T. Keller, G. S. Fonken, J. Szmuszkovicz, and W. S. Johnson, *J. Org. Chem.*, **27**, 1612 (1962).

‡ M. P. Dreyfuss, *J. Org. Chem.*, **28**, 3269 (1963).

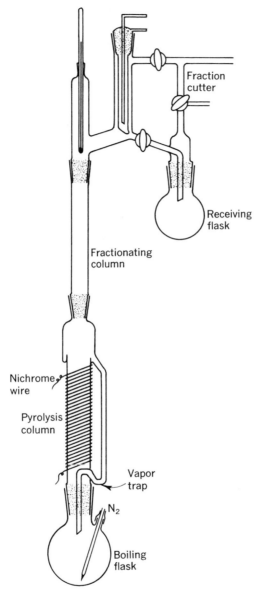

Fraction
cutter

Receiving
flask

Fractionating
column

Nichrome
wire

Pyrolysis
column

Vapor
trap

N_2

Boiling
flask

Fig. 1-18 Recycling pyrolysis apparatus.

When the reaction begins, a solution of allyl bromide in acetone is added dropwise to the flask. In this manner, allyl magnesium bromide reacts with acetone as soon as it forms and complications due to coupling are minimized.

If it is necessary to prepare the Grignard reagent and subsequently utilize it in another reaction as, for example, in the preparation of a carboxylic acid, a Soxhlet reactor† is useful (Fig. 1-19). The reactor is constructed as

† E. Campaigne and O. E. Yokley, *J. Org. Chem.,* **28,** 914 (1963).

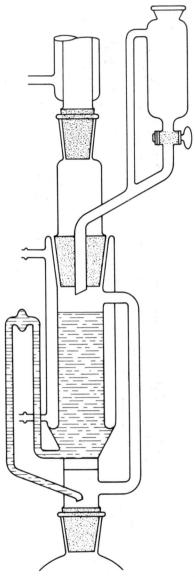

Fig. 1-19 Soxhlet reactor.

shown and the central compartment is filled with magnesium turnings. Anhydrous ether is placed in the round-bottomed flask and the addition funnel and condenser are set in place. The ether is heated to reflux and when the reaction chamber has filled with solvent, slow addition of the allyl bromide from the addition funnel is begun. As allyl bromide is washed down through the reactor by the condensing ether vapors, it reacts with magnesium to form allyl magnesium bromide which passes through the side arm into the round-bottomed flask. The yields of this Grignard reagent are excellent using this method.

1.8.6 Continuous Reactions: As a reaction vessel for a unimolecular reaction such as an ester pyrolysis, one thinks first of a "pyrolysis tube," which is the simplest sort of continuous reactor. There are certain advantages to running *bimolecular* laboratory reactions in a similar manner: Reaction times can be shorter, yields are higher (especially when heat-sensitive substances are involved), and less solvent is required. For large scale operations such as the first reactions in a long multistep synthesis, continuous reactors are worth considering. Two reactions are used to illustrate the technique. In the first, reactants are added from the top; the volatile product distills out, and the nonvolatile product collects at the bottom. In the second, the nonvolatile reactant is added from the top and the volatile reactant from the side; the products collect as before.

The first example involves the reaction of ethyl benzoylacetate with aniline to give benzoylacetanilide and ethanol.[†]

$$\phi COCH_2COOEt + \phi NH_2 \longrightarrow \phi COCH_2CONH\phi + EtOH$$

A mixture of the reactants is added via a dropping funnel [Fig. 1-20(a)] to the heated (135°) column at the rate of 400 g per hr. Ethanol collects in flask B and ethyl benzoylacetate in flask A.

The second example deals with the preparation of an acid chloride. This is usually accomplished by heating the carboxylic acid in refluxing thionyl chloride until the evolution of sulfur dioxide and hydrogen chloride ceases and then purifying the product by distillation or crystallization. This approach is unsatisfactory if the acid chloride—like oleoyl chloride—cannot tolerate prolonged heating. An excellent yield of oleoyl chloride, however, was obtained using the continuous reactor depicted in Fig. 1-20(b).[‡]

$$CH_3(CH_2)_7CH\!\!=\!\!CH(CH_2)_7COOH + SOCl_2 \longrightarrow$$
$$CH_3(CH_2)_7CH\!\!=\!\!CH(CH_2)_7COCl + SO_2 + HCl$$

Thionyl chloride is placed in the three-necked flask and heated to reflux. It distills through the U-tube and column, both of which are heated by electrical tape, is condensed by the Friedrichs condenser, and returns to the original flask. After the rate of distillation of thionyl chloride has reached a steady state, the acid is added dropwise to the column, which serves as a reaction chamber. As the acid chloride forms, it falls into the flask at the bottom of the column, while the gaseous by-products pass from the outlet provided into a gas absorption trap. This technique, based upon a countercurrent reaction principle, produces the acid chloride in yields of 97–99% and can be run on a continuous basis if desired. It is suitable for the

[†] C. F. H. Allen and W. J. Humphlett, *Organic Syntheses,* Coll. Vol. IV, p. 80.
[‡] C. F. H. Allen, J. R. Byers, Jr., and W. J. Humphlett, *Organic Syntheses,* Coll. Vol. IV, p. 739.

preparation of acid chlorides which boil sufficiently higher than thionyl chloride.

1.8.7 Reactions at −30°: Many organic reactions are carried out below room temperature, and most of these pose no special problems. In some cases, however, it is necessary to transfer a very cold solution from one reaction vessel to another at a controlled rate. The preparation of cyclopropyl

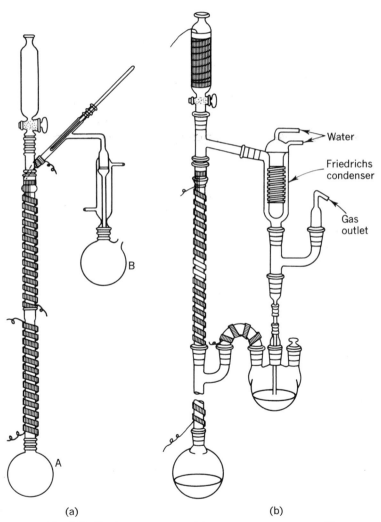

(a)   (b)

Fig. 1-20 Continuous reactors: (a) reactants enter at top; (b) one reactant enters at top, one at side.

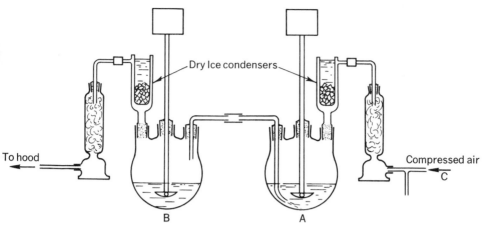

Fig. 1-21 Apparatus for the preparation of cyclopropyl cyanide.

cyanide from γ-chlorobutyronitrile† illustrates a method for accomplishing this.

$$2Na + 2NH_3 \longrightarrow 2NaNH_2 + H_2$$

$$ClCH_2CH_2CH_2CN + NaNH_2 \longrightarrow \triangleright\!\!-CN + NaCl + NH_3$$

The reaction is carried out in two stages, the first of which involves the preparation of sodium amide from liquid ammonia and sodium metal. The apparatus is assembled as shown in Fig. 1-21, and liquid ammonia and a small quantity of ferric nitrate are placed in flask A. Sodium shavings are added to this flask, and it is stirred until the blue color disappears, signaling complete conversion of sodium to sodamide.

In the second flask are placed liquid ammonia and γ-chlorobutyronitrile; while both flasks are stirred vigorously, the sodamide suspension is forced over into flask B by means of air pressure applied through C. The rate of addition is controlled by placing a finger over the bypass in the T-tube. The yields of cyclopropyl cyanide obtained in this way are 52–53%.

† M. J. Schlatter, *Organic Syntheses,* Coll. Vol. III, p. 223.

2
Isolation Techniques

The isolation of a single component from a mixture of compounds is probably the most common problem that an organic chemist encounters. Reaction "workup," which usually requires more time than the actual running of the reaction, is the most frequent isolation problem; the separation of natural products from very complex mixtures obtained from plants or animals is a primary concern of chemists dealing with these substances.

The isolation techniques that are used most frequently are extraction, crystallization, sublimation, distillation, and chromatography; each of these is discussed in some detail. Each technique has advantages and limitations, and it is usual to utilize two or more of them in a particular isolation scheme. The choice of techniques is influenced by the chemical and physical properties of the substances in the mixture, the quantities of materials involved, and the final degree of purity that is desired.

Purity, in a practical sense, is not an absolute state but, rather, a region defined by the ability to detect contaminants with the tools and techniques that are available. The degree of purity desired is governed primarily by the use to be made of the substance being purified. For example, if an intermediate in a multistep synthesis contains an impurity, it would be foolish to expend a considerable amount of time and effort (along with the inevitable sacrifice of some of the intermediate) to remove this impurity if it is chemically inert and would be more easily removed after the next reaction in the sequence. On the other hand, if a reaction product is to be used as a starting material in an accurate kinetic study, painstaking care may be necessary to remove a minor contaminant exerting a catalytic effect on the reaction. Similarly, if a drug is being synthesized for human use, considerable effort must be made to ensure that it is of high quality. In general, the desirability of working with analytically pure compounds must be balanced against the effects of the impurities and the effort required to remove them.

In any purification procedure, it is necessary to have a means for determining the purity of the desired substance in order to recognize when the necessary degree of purity has been reached. One of the most sensitive methods is gas chromatography (Sec. 2.5), which allows quantitative estimates of impurities to be made. The peak for the desired substance can be run far off scale to detect minor impurities with different retention times. For analytical purposes, it is advisable to use more than one column since impurities that have the same retention time as the desired product on one may be separable on another. Thin layer chromatography (Sec. 2.6.5) can be used in a similar manner, but it permits only very crude quantitative estimates. Melting points are quite sensitive to impurities, and if a new substance has been recrystallized to a constant sharp melting point, it is likely to be sufficiently pure for most purposes. Elemental analysis (Sec. 3.2) and NMR spectrometry (Sec. 3.3.1), while not particularly sensitive, can be used to obtain quantitative estimates of certain types of impurities.

Before considering each of the common isolation techniques at length, it is worthwhile to outline the ways in which they are most commonly combined. It should be remembered that while the combinations outlined will suffice for most separations, the best combination for a particular isolation may not be mentioned below.

Simple *extraction* is generally employed first because it is a very rapid and efficient way of removing certain types of impurities. For example, in most organic reactions, the desired product is a water-insoluble solid or liquid, and at the end of the reaction it is in solution. Water and ether or pentane are often added, and the aqueous layer (containing soluble ionic by-products) is separated and discarded. If any by-products are sufficiently acidic to form salts with sodium hydroxide or basic enough to form salts with hydrochloric acid, the organic layer is then washed with the appropriate aqueous solutions, which are discarded. (In cases in which the desired product itself forms a salt with sodium hydroxide or hydrochloric acid, the aqueous solution containing this salt is saved, the organic layer is discarded, and the desired product is liberated from its salt by neutralization of the aqueous layer and is reextracted into ether or pentane.) If the reaction solvent is very soluble in both pentane and water (e.g., ethanol and tetrahydrofuran), an initial distillation to remove the bulk of the solvent is necessary before the extraction procedures above can be used.

The desired product is now in ether or pentane solution, and it is possibly accompanied by other substances with similar acidity properties. The solution is dried over magnesium sulfate,† filtered, and the bulk of the solvent is removed by *distillation*. What is done next depends on whether the desired substance is a liquid or a solid.

If it is a liquid, it may be sufficiently pure for further use, but it will certainly retain some solvent. Most of the time a *distillation* is employed to remove traces of solvent and other impurities with differing boiling points; a careful *fractional distillation* is necessary to remove an impurity boiling

---

† Calcium sulfate has a much greater affinity for water and thus gives a drier solution, but its capacity is much lower (it forms a *hemi*hydrate) and magnesium sulfate is generally used.

within 30° of the desired product. *Preparative gas chromatography* will probably be used instead of fractional distillation if the quantity is small (under 10 g). If all of these have failed to separate an impurity, further possibilities include *liquid chromatography* (with <30 g), *countercurrent distribution* (with <100 g), and conversion to a crystalline derivative, *recrystallization* of this derivative, and regeneration of the original liquid.

If the product is a solid, *recrystallization* is in most cases a highly selective process, and is applied directly upon removal of the ether or pentane. If the desired substance will not crystallize well at this stage, *distillation, sublimation, liquid chromatography* (with <30 g), *preparative gas chromatography* (with <20 g), or *countercurrent distribution* (with <100 g) may effect sufficient purification themselves, or they may be followed by *recrystallization* to give sufficiently pure material. For extreme purity (>99.99% has been claimed in some cases), *zone refining* can be used as a final step.

In the preparation of small ultrapure samples for microanalysis, small amounts of dust, grease, and solvent can cause misleading results, and a method that removes them is usually employed as the last purification step. For solids, the method of choice for the last step is *sublimation,* used after *recrystallization* to constant melting point and/or *preparative gas chromatography.* If the sample decomposes too readily on heating, the first and last of these methods are inapplicable, and recrystallization must suffice. In this situation, the sample is recrystallized using grease-free distilled solvents and glassware washed with such solvents; before the last crystallization, the solution is filtered through sintered glass into a centrifuge tube using the apparatus depicted in Fig. 2-1. After crystals form, the solvent is removed by centrifuging, decanting, air drying, and drying further in a vacuum desiccator overnight.

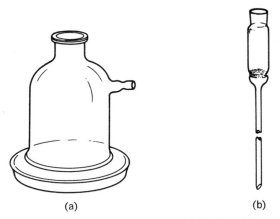

(a) (b)

Fig. 2-1 (a) Micro bell glass (or micro bell jar); (b) micro sintered glass filter funnel.

For analytical samples of liquids, *distillation* is the best final purification step, after preliminary *preparative gas chromatography*. The latter technique is often satisfactory as a last step, but sometimes it gives samples seriously contaminated with column packing and materials of low volatility from previous injections; thus it is safer to follow gas chromatography with a quick evaporative distillation.

The usual container for an analytical sample of either a solid or a liquid is a small screw-cap vial. This should be washed thoroughly with a grease-free distilled solvent, dried in an oven, and the cap lined with a piece of Teflon sheet before use.

2.1 EXTRACTION

Extraction is seldom the sole method used to purify a compound, but it is a rapid and versatile technique that can be used to achieve a preliminary separation prior to a final purification step. Separation of components by extraction depends upon the difference in solubility of a compound in two mutually insoluble phases. Mathematical aspects of extraction are formulated in terms of a simple distribution law, $K = C_a/C_b$, which states that at equilibrium a solute will distribute itself between two immiscible phases, a and b, such that the ratio of concentrations in the two phases is a constant at a given temperature. The constant K is called the *partition* or *distribution coefficient*. If a substance dissolved in solvent b is to be extracted into a second solvent (a), it is obviously advantageous to choose solvent a such that the value of K will be as large as possible. Unfortunately, there is no sure way of predicting K, and the organic chemist relies on the rule that "like dissolves like" and his previous experience in selecting the best solvent system for an efficient extraction.

Suppose, for example, that it is necessary to extract a substrate from an aqueous solution. The first choice that must be made is that of a second solvent. Characteristics that this solvent should have are a low solubility in water, a low boiling point to facilitate eventual removal, and a high solvating capacity for the substance that is being extracted; some commonly used solvents are ether, methylene chloride, and pentane. To decrease the solubility of the organic substrate in the aqueous solution, it is usually good practice to saturate this solution with a salt such as sodium chloride, sodium sulfate, or potassium carbonate. To test the suitability of a particular solvent for the extraction, a measured volume of an aqueous solution of known concentration is extracted once by shaking a measured amount of organic solvent thoroughly with a measured volume of the solvent, and the layers are separated. The organic layer is evaporated and K is calculated from the weight of the residue. From this, the number of simple extractions required for the desired degree of separation is easily obtained. If the number of simple extractions is large, other solvents can be evaluated, or a continuous extraction procedure may suffice.

In a continuous extraction, a small volume of solvent is repeatedly recycled by a distillation-condensation procedure through the solution being extracted. Designs for extractors that can be used to extract an organic substance from an aqueous solution with a solvent less or more dense than the aqueous solution are shown in Figs. 2-2 and 2-3, respectively. In a typical case, an aqueous solution containing the organic substance to be extracted is placed in flask A (Fig. 2-2; a round-bottomed flask can be used here) and covered with an immiscible solvent such as ether. A second flask (B) containing only the organic solvent and a boiling chip is heated to boiling and the solvent is vaporized, liquified by the condenser, and forced to pass

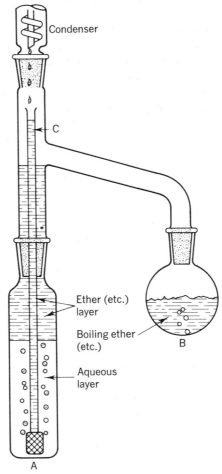

Fig. 2-2 Continuous liquid-liquid extractor for solvents less dense than water.

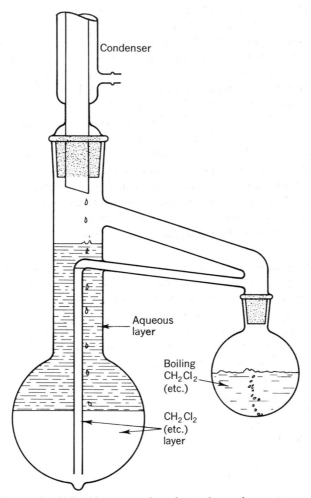

Fig. 2-3 Continuous liquid-liquid extractor for solvents denser than water.

through the aqueous phase by means of tube C. This causes the organic solvent above the aqueous phase to flow over into flask B and maintain the level of solvent in flask B at a constant value. As the process continues, the solute concentration in the aqueous phase is gradually depleted as it concentrates in flask B. A porous glass outlet at the bottom of tube C is advantageous since this serves to disperse the organic solvent into small globules, thereby greatly increasing the contact surface area between the aqueous and organic phases as the organic solvent passes through the aqueous layer; this decreases the time required for a thorough extraction by a significant factor. The apparatus in Fig. 2-3 is used similarly with solvents denser than water. Although a continuous extraction always results in a more efficient separation than a batch process, it is more time consuming and a judgment must be made as to whether or not the increased efficiency will justify the extra time required.

More elaborate multiple extractors based on a countercurrent principle have been developed that are capable of giving very efficient separations in favorable circumstances. For separating amounts greater than about 40 g, a counterdoublecurrent apparatus is preferable. With this apparatus, continuous rather than batchwise operation is possible, if a solvent system can be found such that the distribution coefficient of the desired substance is on one side of unity and those of all impurities are on the other side. Due to the cost of the apparatus (about $8000), the limited quantity of material that can be used (a maximum of about 40 g of substrate), the time necessary to find a suitable solvent system, and the problems associated with running and cleaning the apparatus, countercurrent extractors are usually employed as a last resort, i.e., after distillation, recrystallization, simple extraction, and liquid chromatography have failed to effect a satisfactory separation.

Frequently it is found that there is a substantial difference in acidity and/or basicity between the substance which is to be isolated and the contaminants with which it is associated. Under these circumstances a rapid and efficient separation can usually be achieved through simple extraction, and this should always be considered as the method of choice. Chart 2-1 outlines a general scheme for separating a mixture of acidic (HA), basic (B) and neutral (N) components. The usually valid assumption is made that the organic salts are distributed almost entirely in the aqueous layer and all other organic substances almost entirely in the ether layer.

For the extraction of soluble components from a solid such as bark, two procedures are commonly used. After the solid has been finely ground,

Chart 2-1

SEPARATION OF ACIDIC (HA), BASIC (B), AND NEUTRAL (N) SUBSTANCES
BY AN EXTRACTION PROCEDURE

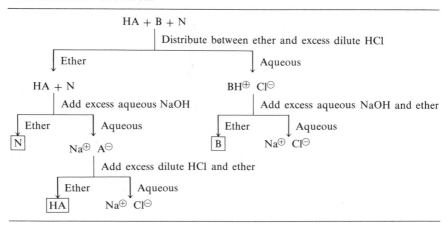

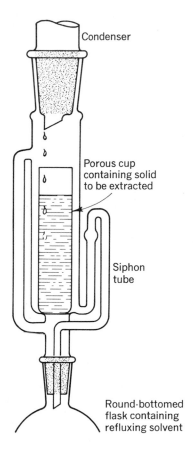

Condenser

Porous cup
containing solid
to be extracted

Siphon
tube

Round-bottomed
flask containing
refluxing solvent

Fig. 2-4 Soxhlet extractor.

generally in a Waring Blendor, it is mixed with a solvent, heated, and filtered, and this process is repeated several times. This procedure is rapid but frequently requires a large total volume of solvent.

A second approach is to place the material to be extracted in a Soxhlet apparatus (Fig. 2-4), which works on a principle similar to the liquid-liquid extractors previously described. The finely ground substrate is placed in a porous paper cup, and the solvent to be used is then heated to boiling, condensed, and dripped into the porous cup. When the level of the solvent rises to slightly above the top of the siphon tube, the solvent in the cup chamber drains back into the round-bottomed flask and the cycle begins again. By varying the solvent polarity, successive extractions can be used to effect a partial separation of a crude substrate into its various components. The Soxhlet method has the advantage of requiring far less solvent and little effort on the part of the chemist, but it is usually considerably slower than the batch extraction described above, and it has the additional disadvantage that the substance being extracted is heated for a long time.

2.2 CRYSTALLIZATION

Purification of solids almost always involves crystallization. Crystallization usually results in a greater loss of material than distillation (Sec. 2.4), but the degree of purity that can be realized is, in general, significantly higher. Further advantages are that crystallization techniques are readily adaptable to both macro and micro scales and that the melting point of each crop of crystals serves as an indication of the degree of purity.

The recrystallization technique most commonly used by organic chemists involves dissolving the solute in a suitable hot solvent, cooling the solution to room temperature or below, and collecting the crystals that deposit on standing. This procedure is much faster than one which depends upon slow evaporation of the solvent from a saturated solution, although the latter method is preferable when high quality single crystals are desired (e.g., for X-ray crystallography).

Suppose that one is faced with the problem of separating and purifying a crystalline compound from a reaction mixture or from natural sources and that the desired compound is contaminated with gross amounts of impurities. An attempt to purify the compound by direct crystallization from the mixture will usually be unsuccessful since the ease of crystallization decreases rapidly as the complexity of the mixture increases. For this reason, a preliminary purification, probably involving one or more of the other methods described in this chapter, may be a necessity.

After the preliminary separation, the next (and most critical) step is the selection of a suitable crystallization solvent. Ideally, the solvent either should not dissolve the impurities at all or should dissolve and retain all of them, while depositing all the desired compound. The desired compound should be much more soluble in hot solvent than in cold solvent. Some commonly used solvents are listed in Table 2-1 in order of decreasing dielectric constant.

To determine which solvent most closely meets these conditions, some small scale experiments are performed. Samples of about 10 mg of the solid (or oil) are placed in small centrifuge tubes, and up to 0.5 ml of a solvent is added dropwise with stirring; if solution is incomplete, the mixture is heated to the boiling point, centrifuged, and the supernatant liquid is cooled (if necessary with stirring and scratching) to induce crystallization. The crystals can be weighed to determine the percentage of recovery. If the substance is completely soluble at room temperature, the experiment is repeated with a drastically reduced volume of solvent, and the solution is chilled if necessary to induce crystallization. Melting points of the crystals can be compared to determine which solvent is most effective in purifying the substance.

If a series of such experiments with pure solvents does not yield a satisfactory recrystallization solvent, the experiments above are repeated using solvent combinations. Solvents for this purpose must of course be

Table 2-1

RECRYSTALLIZATION SOLVENTS

Solvent	Boiling point, °C	Melting point, °C	Dielectric constant at ~25°C
Water	100	0	79
Acetonitrile	80	−46	38
Methanol	65	−98	33
Ethanol	78	−117	24
Acetone	56	−95	21
Methylene chloride	41	−97	9
Acetic acid	118	17	6
Chloroform	61	−64	5
Diethyl ether	35	−116	4
Benzene	80	6	2.3
Dioxane	101	12	2.2
Carbon tetrachloride	77	−23	2.2
Petroleum ether	30–60	—	~2
	60–90		
	90–120		
Cyclohexane	81	7	2.0
Pentane	36	−130	1.8

miscible in the proportions used, and the solvent combination must include at least one member in which the substrate is quite soluble and at least one in which it is quite insoluble.

After suitable solvents or solvent combinations are uncovered in such preliminary studies, further micro scale experiments should be carried out to define the solubility limits of the hot and cold solvents more precisely before the scale of the crystallization is increased. From a combination of these data, successful and well-defined crystallization conditions are found at least 90% of the time.

The size of the crystals formed is largely a function of the rate of crystal growth. The slower a solution is cooled, the larger (and often the purer) the crystals are. Seeding with crystals of the desired substance is always a good idea if crystals are available.

Frequently crystals are well formed and melt sharply but have a colored impurity present. In many cases, if a solution containing the compound is boiled with a small quantity of activated charcoal, the colored impurity is adsorbed preferentially; if this is unsuccessful, a pass of the solution through a short alumina column (Sec. 2.6.1) can be attempted.

After the crystals have been formed, they are collected by suction filtration or decantation and should be washed thoroughly to remove traces of mother liquor. This is done best by suspending the crystals in fresh solvent and refiltering since washing directly on a funnel frequently results in channeling of the liquid through the filter cake and incomplete removal of the mother liquor. If the solvent used is nonvolatile so that its complete removal from the crystals is difficult, it is best to wash the crystals several times with a low boiling solvent in which they are insoluble but with which the crystallizing solvent is miscible.

If the compound is sensitive to air or is to be crystallized at low temperatures, more elaborate filtration equipment (Fig. 2-5) is required. A variation of this design with a jacketed filter through which hot or cold liquids can be circulated is available and is convenient for the recrystallization of low melting substances or substances with inverse solubility curves.

A recrystallization technique called *zone melting* or *zone refining* is occasionally used to bring an organic solid to a very high degree of purity. The solid is placed in a long tube and sufficient heat to melt a small zone of the solid is applied at one end. The heated zone is moved slowly along the tube to the other end and, due to the selectivity of the crystallization process, the impurities are carried along in the liquid zone and concentrated at the other end. Further passes can be made if the desired purity is not attained in one pass.

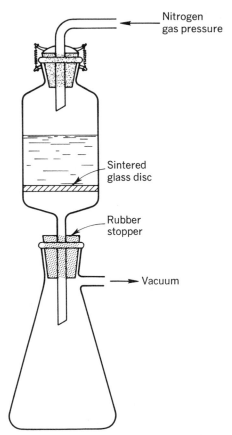

Nitrogen gas pressure

Sintered glass disc

Rubber stopper

Vacuum

Fig. 2-5 Apparatus for the filtration of compounds sensitive to the atmosphere.

2.3 SUBLIMATION

When a solid has a high vapor pressure, purification can occasionally be accomplished by sublimation. This is done by warming the solid to a temperature below its melting point and condensing the vapors on a cold surface. It is frequently practical to go above the melting point since the rate of evaporation is much greater than the rate of sublimation; if there are only nonvolatile impurities present, purification is far more rapid.

Sublimations can be carried out under pressure or vacuum with equal ease. The technique is useful only if impurities associated with the component being sublimed have a substantially different vapor pressure at the sublimation temperature. It is often used as a final purification step in the preparation of an analytical sample. A simple but effective apparatus is shown in Fig. 2-6; the sample is placed in the bottom of the outside tube, it is heated, and crystals of sublimate collect on the large cold-finger condenser. A simple tube such as that illustrated in Fig. 2-7 can be used for

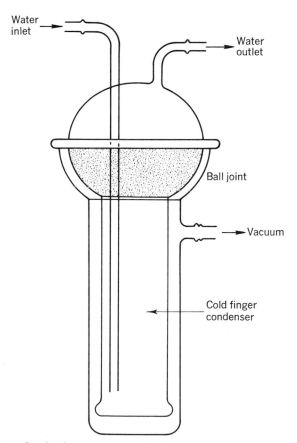

Fig. 2-6 Apparatus for simple or vacuum sublimation.

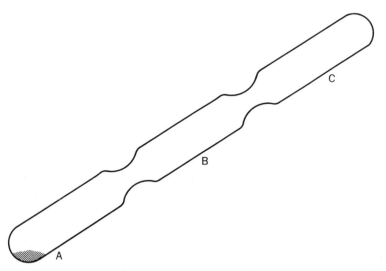

Fig. 2-7 Tube for multiple sublimation.

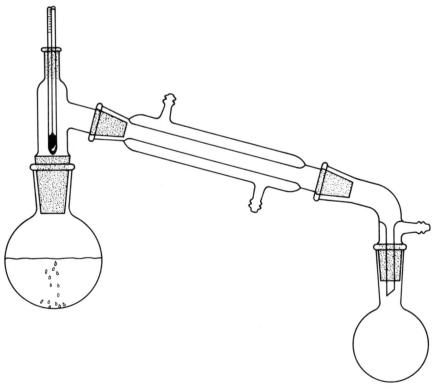

Fig. 2-8 Apparatus for simple distillation.

repeated sublimation without exposing the crystals to the atmosphere between sublimations.

2.4 DISTILLATION

The only practical method of purifying liquids on a large scale is distillation. Equipment useful for simple distillations is shown in Figs. 2-8, 2-9, and 2-10. With the proper equipment, careful operation, and (above all) patience, compounds that have boiling points that differ by only a few degrees can be routinely separated. The components required for fractional distillation are a heat source, distilling and collecting vessels, a column, and a fraction cutter. Of these, the single most important item is the fractionating column since this controls the degree of separation that can be attained.

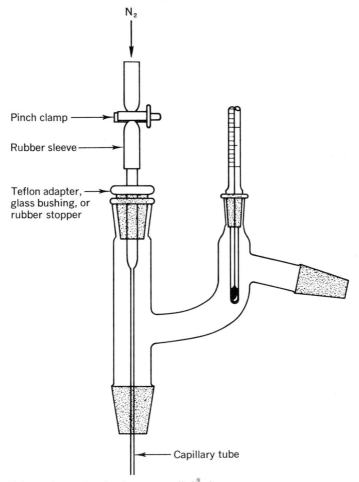

Fig. 2-9 Claisen adapter for simple vacuum distillation.

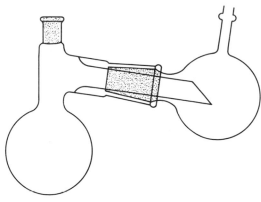

Fig. 2-10 Solid distillation apparatus.

2.4.1 Fractionating Columns: The efficiency of a column is expressed as its number of theoretical plates, which is essentially the number of simple distillations required to effect the same degree of separation that can be achieved by a single distillation through the particular column under specific operating conditions. Since columns are not of uniform length, efficiencies are commonly given in HETP (height equivalent to a theoretical plate) units. The HETP rating of a column is a sensitive variable and is influenced by the rate of distillation through the column and by other operating conditions. A column can work at its rated efficiency only when liquid-vapor equilibrium is established, and this is why skill and patience on the part of the operator is one of the critical factors in determining the success of a distillation.

The selection of a column for a particular distillation is governed by the boiling point differences of the liquids being separated, the volume of the material being distilled, and the degree of separation desired. Table 2-2 contains a list of boiling point differences and the corresponding number of theoretical plates that are required to effect a "good" separation (*ca.* 99% purity of the distillate).

Table 2-2
THEORETICAL PLATES NECESSARY FOR SEPARATION

Boiling point difference, °C	Plates required
30	8
20	13
10	22
7	35
5	50
4	65
3	80
2.5	100

Probably the most widely used column (and the most overrated by students uninitiated in the art of distillation) is the Vigreux column (Fig. 2-11). This consists of a tube with frequent indentations throughout its length that serve to increase the effective surface area of the column. For maximum efficiency, the column should be silvered and vacuum jacketed to minimize external heat losses. If this is not practical, the column can be wrapped in glass wool and bound with asbestos tape to increase its efficiency. A typical HETP value for a Vigreux column is 10 cm. Since it is a rule of thumb that the reflux ratio (the number of drops of liquid returned to the column for each drop collected) should be about equal to the number of theoretical plates at total reflux, a 50-cm Vigreux column operating at a reflux ratio of about 5:1 is capable of delivering about five theoretical plates. This corresponds to an acceptable separation (*ca.* 95% purity of the distillate) when the components of a mixture boil about 30° apart.

Two distinct advantages of Vigreux columns are a low holdup of material and a high throughput rate. These characteristics suit the Vigreux column

Fig. 2-11 Vigreux column with vacuum jacket. The indentations in the column are made by heating a small area of glass tube and making a series of dimples with a pencil-shaped carbon rod.

for semimicro applications and for solvent purification if the impurities have boiling points sufficiently different from the solvent.

An excellent commercially available column in a complete distillation unit is the Todd column (*ca.* $200). This is constructed from a vacuum-jacketed glass tube within which is inserted a metal core that has wire wound about it in a spiral fashion for the entire length of the tube. The 90-cm commercial column operating at a throughput of 1.65 ml per min has been rated at an HETP of 1.8 cm (50 theoretical plates per column) with a holdup of 0.4 ml per theoretical plate. These characteristics make the Todd column suitable for separations of moderate to large quantities of liquids with close boiling points. The commercial unit has the further advantage that numerous packed columns can be interchanged with the spiral column.

Probably the best general purpose column for laboratory fractional distillation is the spinning-band column (Fig. 2-12).† This consists of a straight vertical tube that contains a platinum or stainless steel wire or Teflon gauze with diameter very slightly less than that of the tube that extends the entire length of the tube. During the distillation, this wire band is spun rapidly (*ca.* 3000 rpm), creating turbulence and increasing the diffusion coefficient of the distilling vapor. For a typical 75-cm column, 1 cm in diameter, operating at a throughput of vapor corresponding to 3.4 ml of liquid per minute, the HETP is 1.9 cm (40 plates per column) with a holdup of 0.018 ml per plate. The combination of low holdup and high efficiency makes this column suitable for the high efficiency distillation of small or large quantities of liquids. The negligible pressure drop through the column is especially desirable for vacuum distillations.

Another column excellent for fractionating small quantities of liquids at atmospheric pressure is the concentric tube column. This column is constructed from two precision bore tubes mounted concentrically with a very small annular space between them. During the distillation, the vapors and liquid move through this annular space and fractionation occurs. These columns are characterized by low throughput rate and a very low HETP and holdup. Some typical data for a 30.5-cm column, 6.5 mm in diameter, operated at a throughput rate of 1.2 ml per min, are an HETP of 0.36 cm (85 plates per column) with a holdup of only 0.018 ml per plate. These characteristics make this column ideal for the fractional distillation of a few milliliters of a liquid or for the purification of an analytical sample.

In addition to the aforementioned columns, there is a large family of packed columns that merit consideration for use in fractional distillations. These consist of jacketed tubes filled with packing that varies from glass

---

† Available for *ca.* $600 from Nester/Faust Mfg. Corp., 2401 Ogletown Rd., Newark, Del. 19711.

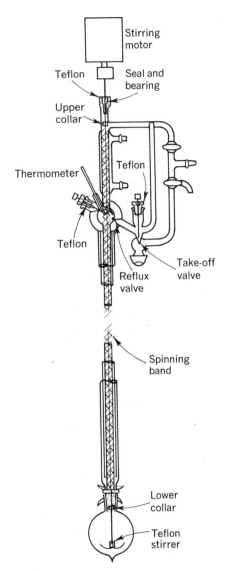

Fig. 2-12 Spinning-band column.

beads and helices to metal grids. These are excellent columns for the distillation of large quantities at atmospheric pressure and are characterized by moderate holdup and high efficiencies. Unless great care is taken, they have a pronounced tendency to flood during operation and this is a distinct disadvantage, especially if the operator is unskilled in the art of distillation. Directions for building such a column accompany Fig. 2-13.

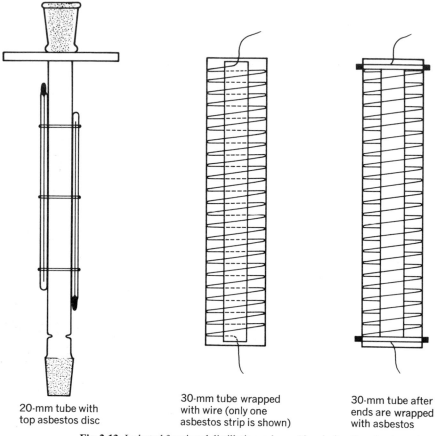

20-mm tube with
top asbestos disc

30-mm tube wrapped
with wire (only one
asbestos strip is shown)

30-mm tube after
ends are wrapped
with asbestos

Fig. 2-13 Jacketed fractional distillation column (for clarity, the asbestos strips have been omitted).

The directions given apply to a 2-ft column but can be modified as desired.

1. Construct the fractionating column from 20-mm Pyrex tubing such that the distance between the top flange and the constriction at the base of the tube is 60 cm. Cut a piece of 30-mm tubing 60.8 cm long and a piece of 48-mm tubing 60.4 cm long.

2. From $\frac{1}{4}$-in. asbestos board, cut out two matching discs as shown, making the center hole just large enough to accommodate the fractionating column. Slip one disc on the fractionating column and affix the thermometers in place with fine Nichrome wire using a small amount of wet asbestos paper to prevent slippage.

3. Cut out four (or six) strips of asbestos paper about 1 cm in width and 60 cm long, moisten them with water containing a little water glass, and press them against the 30-mm tube so that they are equally spaced. After drying in an oven, wrap the 30-mm tube with high resistance Nichrome wire (No. 24 is generally satisfactory) in a spiral fashion, spacing each turn about 1 cm apart for the length of the column, and place a second strip of wet asbestos paper over the wire and original strip. A few layers of asbestos paper wound around the bottom and top of the jacket about 5 mm from each end prevents the wire from unwinding after

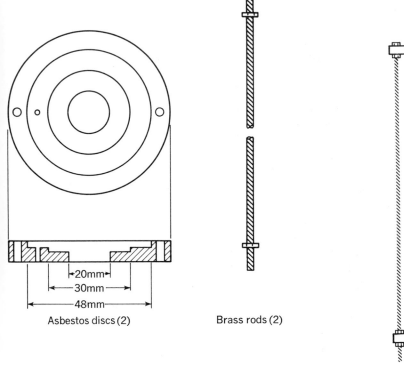

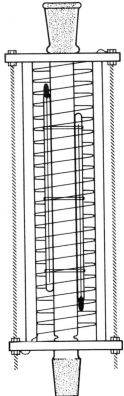

20mm
30mm
48mm

Asbestos discs (2)   Brass rods (2)

Fig. 2-13 (cont.)

the paper has dried. To prevent the wire from unraveling during the winding process, it may help to hold the wire in place with small strips of adhesive tape at strategic intervals. About 20 ft of wire is required for a 2-ft column.

4. Cut two 62-cm lengths of $\frac{1}{8}$-in. brass rod and thread about 1 in. of each end. Thread a nut on each end of both rods.

5. Clamp the fractionating tube on a ring stand upside down and slide the heating jacket over it, slipping the heating wire through the hole provided for it. Place the outer jacket in position along with the two brass rods, and place the remaining disc in its proper position; clamp the top end of the tube to the ring stand. Adjust the position of the nuts to the correct spacing between the top and bottom discs to assure a snug but strain-free fit for the entire assembly and then bolt down the outer faces of each disc using four additional nuts, attaching each lead wire from the heater to a *different* brass rod. The column is now complete.

6. To operate the heater, attach the leads from the wires plugged into a Variac and fitted with alligator clamps to the brass rods. Gradually vary the applied voltage, allow the column to equilibrate, and record the temperature of the column. Retain this calibration for use during distillations.

7. Pack the column carefully with any suitable material. Some common packings in order of decreasing efficiency are Heli-Grids, Heli-Pak, glass helices, and glass beads.

Numerous other columns have also been devised and used. For more detailed information on the characteristics of the columns above and others not mentioned in this text, the excellent treatise on distillation edited by Weissberger† should be consulted.

Since any effective distillation requires that a close to true equilibrium exists between the boiling liquid and condensing vapors, it is important to minimize external heat losses in the system. This is most easily accomplished by enclosing the fractionating tube in a vacuum jacket and placing the jacketed tube in a second jacket that can be heated electrically. After the liquid is refluxing steadily in the column, the external jacket temperature should be adjusted to about 1–2° below the reflux temperature. If the column floods at the bottom, either the boil-up rate is too great or the column jacket is too cold. If the column floods at the top, the column jacket is probably too hot. Another common source of flooding with packed columns is the presence of a constriction due either to improper packing or to a rearrangement of packing caused by bumping or a prior flooding condition. If there is a constriction, the column must be disassembled and repacked.

2.4.2 Heating: The two most common heat sources for distillation flasks are oil baths and heating mantles, both described in Sec. 1.7.1. The heat source should allow fine adjustment of the heat input rate and ideally should apply the heat uniformly to avoid local superheating with resultant decomposition (more of a problem with heating mantles than with oil baths).

A problem related to boiling the liquid is the "bumping" of the charge being distilled. Since bumping seriously affects the equilibrium in a column, it is intolerable in any distillation. In operations at atmospheric pressure, almost any inert boiling chips will prevent bumping, but these are generally unsatisfactory for use during vacuum distillations. The only method certain to assure smooth ebullition during a distillation at reduced pressure is to provide a gas bleed through a very fine capillary tube extending directly into the liquid that is being distilled [Fig. 2-14(a)].‡ Air is employed if the compound is insensitive to oxygen at the pot temperature; otherwise, nitrogen directly from a cylinder or from a balloon is used. The rate of gas flow, which should be minimal for obvious reasons, can be controlled to the point where bubbles are barely visible by attaching a rubber sleeve to the end of the capillary tube and varying the size of the orifice through the tube with a screw clamp. Rapid stirring of the refluxing liquid by means of a magnetically driven stirring bar has also been used to avoid bumping, with limited success.

2.4.3 Distilling Flasks: Distilling flasks are available in many sizes and shapes and are frequently designed for a specific type of column. If a choice

---

† *Technique of Organic Chemistry,* Vol. IV.

‡ An exception to this generalization is provided by spinning-band columns (Fig. 2-12); these have a *very* rapidly rotated Teflon stirrer that works under vacuum.

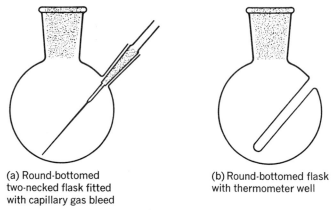

(a) Round-bottomed
two-necked flask fitted
with capillary gas bleed

(b) Round-bottomed flask
with thermometer well

Fig. 2-14 Special distilling flasks.

is available, pear-shaped flasks [Fig. 1-1(b)] are generally preferable for the distillation of small quantities of liquids since there is less holdup in the distilling pot. Some flasks [Fig. 2-14(b)] have a thermometer well built in, and these are advantageous if a record of the temperature of the refluxing liquid is desired. The well should be partially filled with mineral oil to assure a more accurate thermometer reading. The fewer the number of necks in a distilling flask, the smaller the likelihood of leaks in the system; but two necks are usually desirable since it is frequently necessary to add or remove liquid during the distillation or to provide for a gas bleed into the system. None of the considerations above is really critical and the selection of a distilling flask is largely a matter of personal preference.

Figure 2-15 shows a micro-Hickman still, which is constructed from any size glass tubing by first blowing two bubbles as shown in part (a) and then collapsing the upper bubble to the shape shown in part (b). These stills are extremely useful for the purification of small amounts of liquids for analysis, etc., by *evaporative distillation.* The number of theoretical plates is very low, but low boiling solvent and high boiling material such as stopcock grease and dust can be rapidly removed in this way. The sample is placed in the lower bulb (no boiling chip!); if an appreciable amount of low boiling solvent such as ether is present or is to be used to obtain a quantitative transference to the still, the lower bulb should be immersed in an oil bath about 30° above the boiling point of the solvent and the solution added dropwise rapidly. The solvent boils away almost as rapidly as it is added. If atmospheric pressure distillation is to be used, the oil bath temperature is then simply raised to about 10° below the boiling point of the sample, and the sample evaporates and condenses in the upper chamber in a few minutes. If vacuum is to be used, the still is cooled after the low boiling solvent is essentially gone, the vacuum hose is connected to the still, vacuum is applied, and the still is then warmed in the oil bath to 10° below the sample boiling

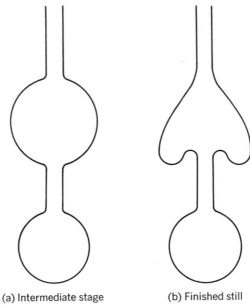

(a) Intermediate stage   (b) Finished still

Fig. 2-15 Micro-Hickman still.

point. The distillate is removed with a capillary medicine dropper with a bent tip; if desired, the still can be centrifuged first to bring droplets of condensate down to the lip of the upper chamber.

2.4.4 Fraction Collection: A component of a fractional distillation assembly critical for the satisfactory operation of the fractionation column is the fraction cutter. Its purpose is to allow collection of the distillate and to control the reflux ratio. A fraction cutter should be designed so that no major disturbance of column equilibrium occurs when a change of receivers is made. A widely used manually operated fraction cutter is shown in Fig. 2-16; this is a typical setup for fractional distillation. Two alternative fraction cutters are shown in Figs. 2-17 and 2-18.

For high precision distillations, it is advantageous to have an automatically controlled fraction collecter. Although these are more expensive, they allow precise control of the reflux ratio and require less attention than their manually operated counterparts. They usually consist of a timing device and a solenoid-controlled valve that is opened automatically on occasion to collect some distillate. The reflux ratio is controlled by the ratio of time the valve spends in the closed and opened positions. In view of the time saved and the greater control available from the use of an automatically controlled fraction cutter, their purchase ($100) is highly recommended.

2.4.5 Vacuum Systems: Since many organic liquids boil at elevated temperatures and are subject to thermal decomposition, it is very often necessary to distill at reduced pressures. Although a moderate vacuum (5–50

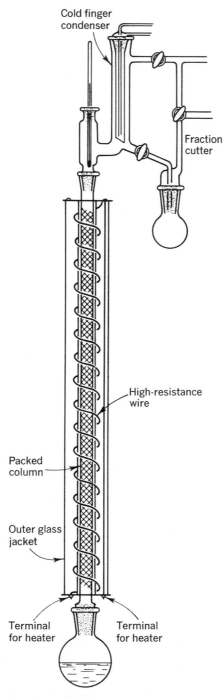

Fig. 2-16 Apparatus for fractional distillation.

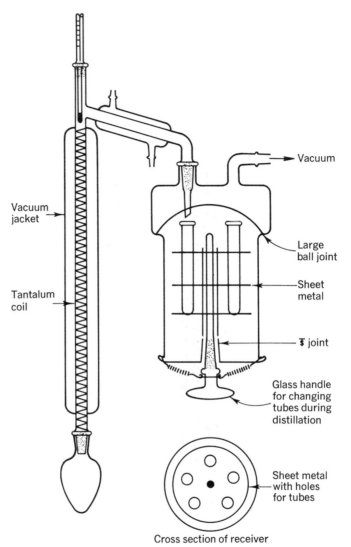

Vacuum

Vacuum
jacket

Tantalum
coil

Large
ball joint

Sheet
metal

⊥ joint

Glass handle
for changing
tubes during
distillation

Sheet metal
with holes
for tubes

Cross section of receiver

Fig. 2-17 Apparatus for fractional distillation of small samples.

mm, i.e., the vapor pressure of water at its temperature; this can be obtained from tables in chemical handbooks) can be achieved with a water aspirator, maintenance and control of the pressure is a serious problem and, as a result, mechanical pumps are usually more satisfactory. For maximum versatility, it is advisable to construct a vacuum system on a movable cart; a suitable arrangement is pictured in Fig. 2-19.

With most vacuum pumps, the limiting pressure which can be reached corresponds to the vapor pressure of the pump oil. To avoid contamination

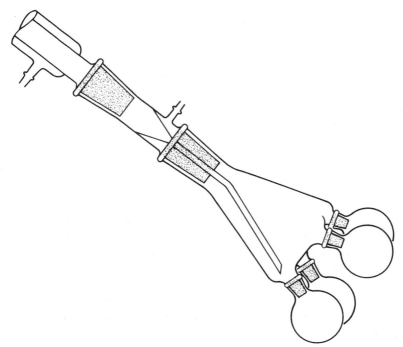

Fig. 2-18 Cow fraction cutter.

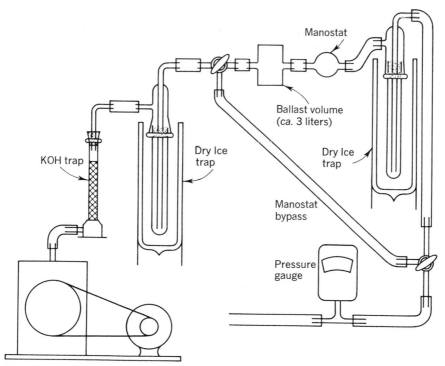

Fig. 2-19 Vacuum system layout.

of the pump oil with organic vapors and corrosion of the pump itself from acidic vapors, it is necessary to protect the pump with appropriate traps. Therefore, immediately before the pump there should be a potassium hydroxide trap and a Dry Ice trap. Since the major cause of breakage of Dry Ice traps is mechanical rather than thermal shock, it is far more economical to use large Thermos flasks (purchased at a drug store for $2) rather than Dewar flasks (purchased from a chemical supply house for $20). The Thermos flask should contain a slush of Dry Ice and acetone or isopropyl alcohol (the latter is preferable since it is less volatile, less flammable, and does not "foam" to the same extent) to condense any volatile vapors. Volatile substances must be removed from the traps before a good vacuum can be obtained. When a solvent boiling below 150° at 1 atm is being removed from a liquid prior to a distillation, it is good practice to remove the remaining traces of the solvent with an aspirator rather than with a vacuum pump unless there is a pump set aside in the laboratory specifically for this purpose.

In addition to these considerations, it is necessary to make provisions for controlling and measuring the vacuum produced by the pump; this requires that a manostat and a manometer be incorporated into the system. Excellent Cartesian diver manostats are available commercially at reasonable prices (*ca.* $50), and these and the Lewis manostat (Fig. 2-20) are very satisfactory for most organic distillations since pressure variations within the system will be very small during their operation. It is desirable to protect the manostat from the system with a Dry Ice trap to avoid fouling of the mercury by contact with organic vapors.

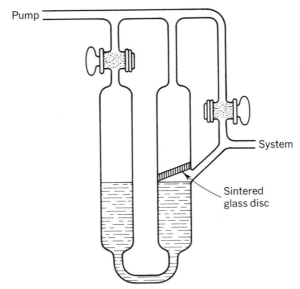

Fig. 2-20 Lewis manostat.

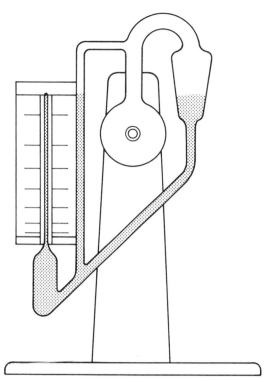

Fig. 2-21 McLeod gauge for measurement of low pressure.

In using a Lewis manostat, the large tubes are half-filled with mercury, and both stopcocks are opened. When the pressure is slightly above that desired, they are both closed, and the manostat controls the pressure in the system by opening and closing it to the pump as required.

Several types of commercial manometers find common use. For pressures below 10 mm, a tilting McLeod gauge (Fig. 2-21) is usually used. It is filled with mercury to the indicated level, and when the pressure is to be measured, it is slowly tilted 90° clockwise. After several minutes to allow pressure equilibration, it is returned to its original position and the pressure is read directly. For pressures from 10–200 mm, the Bennert gauge ($30; shown in Fig. 2-22) is accurate and easy to operate; the pressure in millimeters of mercury is the mercury height difference between the two sides. Universal vacuum gauges give accurate readings from 1 μ to 25 mm but take a long time to equilibrate between readings and are, therefore, not as suitable for most applications in organic chemistry as the types mentioned above.

To assure that the system achieves and maintains as high a vacuum as possible, it is important to use good quality, heavy-walled rubber tubing wired securely to all glass tubing. It is desirable to paint all connections with a preparation such as Glyptal to further reduce the possibility of leaks.

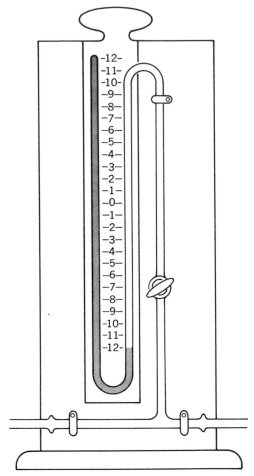

Fig. 2-22 Bennert gauge for measurement of moderate pressure.

All stopcocks should be carefully greased with a high quality vacuum grease, checked frequently for leaks, and replaced if necessary. Tubing should be of reasonably large diameter to facilitate the flow of gases, and a large (*ca.* 3 liter) gas ballast wrapped with friction tape should be incorporated into the system to counteract any sudden pressure changes that may occur, e.g., during the changing of receiving flasks. A careful watch should be kept on the condition of the pump oil, and it is good practice to change the oil and rinse out the oil chamber at least twice a year.

Most laboratory pumps are capable of pulling 0.1 mm or better on a distillation system. If the pressure is higher than this, over 99% of the time it is because the traps need cleaning or the system leaks—often around a poorly fitting or poorly greased stopcock. Convincing evidence that the pump is still effective can be obtained by connecting the vacuum gauge directly to the pump and noting the pressure.

2.4.6 Solvent Removal: After extraction, chromatography, or completion of a reaction, it is frequently necessary to remove a solvent prior to further treatment of the product. An efficient method for accomplishing this is to use a solvent stripper as shown in Fig. 2-23. The sample to be concentrated is placed in a flask that is filled no more than half full, and a fine capillary tube (air or nitrogen bleed) is inserted through the distillation head to the bottom of the distilling flask. The solvent is heated with either a hot water or steam bath while aspirator vacuum is applied. If the rate of distillation

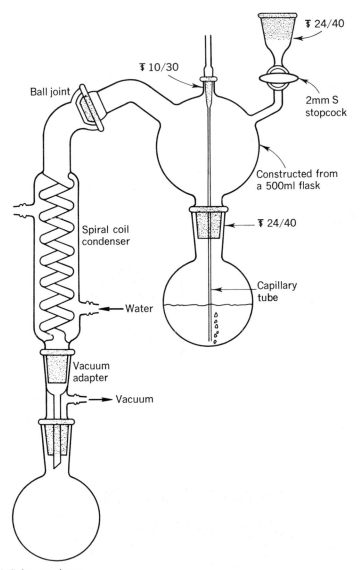

Fig. 2-23 Solvent stripper.

of solvent becomes too rapid, a little air or nitrogen can be let into the system from the stopcock attached to the distilling head; this can also be used to introduce more material if a large quantity of solvent is to be evaporated.

Several commercial rotating evaporators (Fig. 2-24) are also available ($100–600) for rapid solvent removal, and these are generally quite efficient unless they leak around the ring seals.

2.5 GAS CHROMATOGRAPHY

The separation of compounds by passing them into a column packed with a "stationary phase" and then passing a "mobile phase" through the column until the components of the mixture are selectively eluted is called *chromatography*. The chromatographic techniques that employ a liquid mobile phase are discussed in Sec. 2.6. The most widely used chromatographic technique, discussed in this section, employs a gaseous mobile phase and is variously called *gas chromatography* (GC), *vapor phase chromatography* (VPC), *gas-liquid chromatography* (GLC), and *gas-liquid phase chromatography* (GLPC). In the last two of these names, *liquid* refers to a liquid adsorbed on a solid support in the stationary phase. It might seem that gas chromatography would be limited only to gases or very low boiling

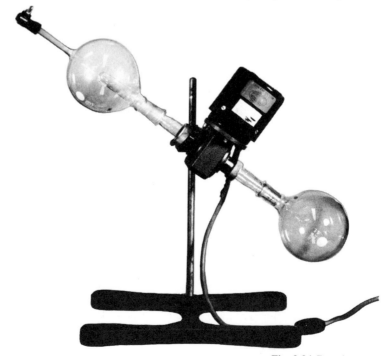

Fig. 2-24 Rotating evaporator.

substances since the substance must be vaporized, but actually a great many organic liquids and solids are sufficiently volatile at *ca.* 300°, the maximum temperature commonly employed for this technique. Many polar compounds such as sugars that are not volatile enough for GC directly can be run as trimethylsilyl ether derivatives, easily prepared by exchange with a reagent called "Tri-Sil."†

The rate of elution of a component of the mixture is determined by the relative affinities of the component for the stationary and mobile phases. In gas chromatography, the elution rate is most often expressed as a *retention time,* defined as the time between the placement of the sample on the column and the maximum elution of the component in question. The basic principle responsible for the great utility of all chromatographic methods is the same: Even compounds with very similar structural features frequently are eluted from the column at sufficiently different rates that an analytical or preparative separation is possible. If the efficiency of a gas chromatographic column is expressed in theoretical plates for comparison with distillation (Sec. 2.4.1), it is found that the gas chromatographic column provides *50 times* as many theoretical plates as an efficient distillation column (e.g., a spinning-band column) of the same length.

After a brief discussion of a gas chromatograph and its use in Sec. 2.5.1, more detailed instructions concerning some of the key parts of the system are given in the later sections.

2.5.1 Operation: With a typical gas chromatograph, illustrated in Fig. 2-25 and shown schematically in Fig. 2-26, the flow of carrier gas (helium, if a thermal conductivity detector is being used; helium or nitrogen if a flame ionization detector is used) is initiated and regulated by means of a pressure gauge-pressure reducing valve to a rate of *ca.* 50–100 ml per min. The heaters for the oven and the injection chamber are turned on; the injection chamber usually has a separate heater so that it can be made considerably hotter than the column and will volatilize the sample very rapidly upon injection. It is important that the carrier gas be flowing while the column is being heated since a column can easily be ruined by oxidation if this is not done. If the oven requires a prolonged period of time to reach temperature equilibrium, gas can be conserved by keeping the flow rate at a minimum during the heating period. The recorder is turned on.

After the oven temperature has stabilized, a sample of a volatile solvent such as ether is injected with a syringe‡ into the injection chamber, where it vaporizes and is swept onto the column by the carrier gas. At the end of the column, a detector connected to a recorder signals the emergence of the solvent from the column as a peak on chart paper. If the solvent gives a strong peak within a minute or two as it should, it can be assumed

† Pierce Chemical Co., Box 117, Rockford, Ill. 61105; a brochure on the use of this reagent can be requested with the order.

‡ Hamilton micro syringes, available from the major supply houses, are widely used.

Fig. 2-25 A typical gas chromatograph (*courtesy of Varian Aerograph Co., Walnut Creek, Calif.*).

that the equipment is working properly. If one component of the mixture is of particular interest and an authentic sample of it is on hand, it is injected next to learn its retention time under these operating conditions. If its retention time is found to be inconveniently long, the oven temperature is raised; or if it comes off the column so quickly that other components of the mixture will probably not be well separated from it, the oven tempera-

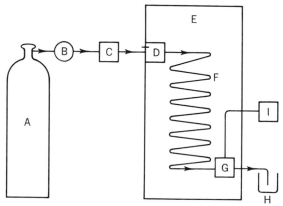

Fig. 2-26 Schematic diagram of a typical gas chromatograph: (A) gas cylinder; (B) pressure gauge and pressure reducing valve; (C) flow meter; (D) injection chamber; (E) insulated oven; (F) chromatography column; (G) detector; (H) sample collector; (I) recorder.

ture is accordingly lowered. The syringe is cleaned of the standard substance by rinsing *about 10 times* with solvent, and a sample of the mixture is injected.

A recorded plot from a typical experiment involving a mixture of three components might appear as shown in Fig. 2-27. The area under a peak is in many cases roughly proportional to the concentration of the component giving rise to it (to obtain accurate compositions of mixtures, known mixtures must be run), and the retention time is a characteristic of its chemical identity. If the aim of the experiment is to achieve a physical separation of the components, they can be condensed and collected after they have passed through the detector (Sec. 2.5.4).

2.5.2 Columns: To realize the full potential of gas chromatography, it is necessary to develop an understanding of the factors that influence the efficiency of a separation by this technique. By far the most important component of the gas chromatograph in this respect is the column. By making an appropriate choice as to the packing and exercising care in the column's construction (if it is not obtained commercially), very difficult separations can be achieved.

The vast majority of column packings consist of a high boiling liquid adsorbed on a solid support with high surface area. The common solid supports, firebrick and diatomaceous earth (sold under a variety of trade names), can be purchased unwashed, acid washed, base washed, and tri-methylsilylated. Solid supports in the first two of these conditions should be avoided if the compounds being investigated are acid sensitive. To permit rapid flow of gas through the column, the solid support should be of fairly uniform particle size; commercial 30–60 mesh granules are usually used, but the adjacent ranges are also suitable.

The critical component of the column packing is the high boiling liquid since the relative retention times of the components of a mixture are governed largely by their relative volatilities from this liquid. The volatility of a component depends on the strength of its interactions with the liquid; the interactions may involve hydrogen bond formation, dipole-dipole association, or complex formation.

The liquid phase should be chosen to take advantage of structural differences between the components of the mixture. In some cases, a column

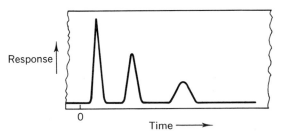

Fig. 2-27 Gas chromatogram of a well-separated three component mixture.

can be selected with considerable confidence a priori that it will work. For example, nonpolar liquids such as silicone oil interact weakly with dissolved nonpolar substances, and the retention times of such substances approximately parallel their boiling points. As another example, a mixture of cycloheptane (boiling point, 118.5°) and cycloheptene (boiling point 115°) would be very well separated using a solution of silver nitrate in a glycol as the liquid phase since olefins form complexes with silver ions and are retained far longer on such a column than alkanes of similar molecular weight. In most cases, however, the interactions between liquids and dissolved substances are not understood well enough for prediction of the order of elution of similar substances, and trial and error must be used in the selection of a column. Most laboratories have a wide variety of columns that can be rapidly interchanged for trial experiments. It should be pointed out that separation can be increased somewhat by lowering the oven temperature or preparing a longer column, but if a significantly greater separation is needed, it is advisable to use a different column packing. A partial list of commonly used liquid phases, maximum operating temperatures, solvents, and areas of useful application is presented in Table 2-3. More complete lists and appropriate literature references are available.†

For analytical purposes, a *capillary* column consisting of a glass capillary tube whose walls have been coated with a stationary phase is sometimes used. It is possible to produce columns of great length (columns 1 mile long have been described) with remarkably high efficiencies that make them ideally suited for very difficult separations. They are harder to prepare than packed columns, however, and during operation, care must be taken not to clog them or exceed their very limited capacity by injecting too much sample.

It is also possible to use the support in a packed column without coating it. Firebrick tends to be acidic and reacts with basic substrates. Other materials that have been used without coatings are Teflon, nylon, graphite, micro glass beads, and carborundum. There appears to be little advantage to supports without coatings, and they are not widely used.

Although columns can be purchased commercially, they are often constructed in the laboratory since there is a substantial saving in cost and an increase in the control of variables (column length, diameter, type and concentration of packing) for application to specific research problems. Packed columns useful for analytical and preparative work are usually 1–50 ft in length and $\frac{1}{8}$–$\frac{1}{2}$ in. in diameter. For short columns, glass can be used but, under normal circumstances, copper, aluminum, or stainless steel tubing is preferable. Of these metals, copper is the cheapest and most flexible, and

---

† See *Guide to Stationary Phases for Gas Chromatography*, Analabs, Inc., Hamden, Conn., and the listings of such companies as Varian Aerograph, 2700 Mitchell Drive, Walnut Creek, Calif. 94598, F & M Scientific Corp., Rt. 41 and Starr Rd., Avondale, Pa., and the Perkin-Elmer Corp., Norwalk, Conn.

Table 2-3

STATIONARY PHASES FOR GAS CHROMATOGRAPHY

Key to solvents used in preparation of column packing

A Acetone	E Ethyl acetate	T Toluene
B Benzene	F Formic acid	W Water
C Chloroform	M Methanol	X Xylene

Phase	Used for	Maximum temperature, °C	Solvent
Alkyl aryl sulfonate (Tide washed with n-C7)		225	W
Apiezon L	Alcohols	300	B
	Aldehydes and ketones		
	Aromatics		
	Boron compounds		
	Drugs and alkaloids		
	Fatty acids		
	Halogenated compounds		
	Hydrocarbons		
	Metals and inorganic compounds		
	Nitrogen compounds		
	Pesticides		
	Pyrolysis		
	Sugars		
	Sulfur compounds		
Apiezon M	Alcohols	275	B
	Essential oils		
	Hydrocarbons		
	Metals and inorganic compounds		
	Sulfur compounds		
Apiezon N	Aromatics	325	B
	Halogenated compounds		
	Vitamins		
Beeswax	Essential oils	200	C
Bentone 34 (organic-aluminum silicate derivative)	Aromatics	200	T
Benzyl cyanide-silver nitrate (2:1)	Hydrocarbons	25	A
Carbowax 200	Aldehydes and ketones	100	C
Carbowax 400	Alcohols	125	C
	Essential oils		
	Nitrogen compounds		
Carbowax 600	Alcohols	125	C
	Amino acids		
Carbowax 750		125	C
Carbowax 1000	Aldehydes and ketones	125	C
	Halogenated compounds		
Carbowax 1000 monostearate		150	C

Table 2-3

STATIONARY PHASES FOR GAS CHROMATOGRAPHY (CONT.)

Phase	Used for	Maximum temperature, °C	Solvent
Carbowax 1500	Alcohols Aldehydes and ketones Halogenated compounds Nitrogen compounds	200	C
Carbowax 1540	Alcohols Essential oils Hydrocarbons Pyrolysis Sulfur compounds	200	C
Carbowax 1500 monostearate		200	C
Carbowax 4000	Aromatics Essential oils Halogenated compounds Sugars	200	C
Carbowax 4000 dioleate		220	C
Carbowax 4000 monostearate		220	C
Carbowax 6000	Sugars	200	C
Carbowax 20M	Alcohols Aldehydes and ketones Aromatics Drugs and alkaloids Essential oils Fatty acids Gases Halogenated compounds Nitrogen compounds Pesticides Phosphorus compounds Pyrolysis	250	C
Dioctyl phthalate (DOP)	Aromatics Halogenated compounds Hydrocarbons Sulfur compounds	140	C
Dow Corning high vacuum grease (methyl)	Essential oils Pesticides Phosphorus compounds Sulfur compounds	350	E
Dow Corning high vacuum grease (ethyl acetate extract)	Pesticides	350	E
Dow Corning Silicone 11 (grease)	Fatty acids Metals and inorganic compounds Pesticides Phosphorus compounds Sugars	300	E
Dow Corning Silicone Oil 200 (12,500 cs) (methyl) or (2,500,000 cs)	Essential oils Fatty acids Hydrocarbons Pesticides Sulfur compounds	250	T

Table 2-3

Phase	Used for	Maximum temperature, °C	Solvent
Dow Corning Silicone Gum 410 (methyl)		325	T
Dow Corning Silicone Oil 550 (methyl phenyl)	Amino acids Aromatics Fatty acids Gases Gases (blood and respiratory) Halogenated compounds Metals and inorganic compounds Nitrogen compounds	275	A
Dow Corning Silicone Oil 555 (methyl phenyl)		275	C
Dow Corning Silicone Oil 703 (methyl phenyl)	Halogenated compounds Hydrocarbons	225	C
Dow Corning Silicone Oil 710 (methyl phenyl)	Aromatics Hydrocarbons Nitrogen compounds Pesticides	300	C
β,β'-Oxydipropionitrile	Alcohols Halogenated compounds Hydrocarbons	100	M
Paraffin wax	Halogenated compounds Hydrocarbons Metals and inorganic compounds	200	C
Silicone gum rubber SE30 (methyl)	Alcohols Aldehydes and ketones Aromatics Bile and urinary compounds Drugs and alkaloids Essential oils Fatty acids Gases Gases (blood and respiratory) Halogenated compounds Hydrocarbons Metals and inorganic compounds Nitrogen compounds Pesticides Phosphorus compounds Steroids Sugars Sulfur compounds Vitamins	350	T

it is usually used unless it reacts with the substances under study.

A column packing is prepared by dissolving an appropriate weight (usually 5–25% of the weight of the solid support) of the liquid phase in a volatile solvent such as pentane, ether, acetone, or methanol. This solution is mixed with the solid support, and the solvent is removed by gentle heating under conditions of constant agitation. A rotary evaporator is ideally suited for this purpose.

A straight section of copper tubing of appropriate length and diameter is plugged at one end with a wad of glass wool, and a funnel to facilitate addition of the packing to the column is connected to the opposite end by means of flexible rubber tubing that fits over both the column and the stem of the funnel (Fig. 2-28). The packing is then added to the funnel in small quantities and the column is tapped gently or vibrated at regular intervals to assure uniform packing. With large columns, care must be taken to avoid the formation of channels and separation of the packing on the basis of particle size; the latter problem can be minimized, at some extra cost, by using a support that has a narrower range of particle size.

When the column is completely filled, the funnel is removed and the open end is packed with a small wad of glass wool to retain the packing. The column is now adapted to the shape of the oven, usually by coiling it in a spiral shape around a cylinder of appropriate diameter. Fittings are added to connect the column to the appropriate terminals in the oven. The

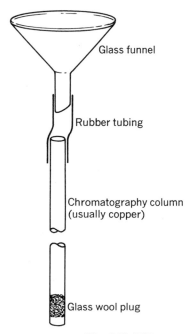

Glass funnel

Rubber tubing

Chromatography column (usually copper)

Glass wool plug

Fig. 2-28 Filling a gas chromatography column.

column is then conditioned by heating it in the oven under a gentle flow of carrier gas for several hours.

If the recommended maximum temperature for the liquid phase is not exceeded for more than short periods, if sample sizes are not excessive, and if acidic substances are avoided, most gas chromatography columns still perform well after thousands of injections.

2.5.3 Detectors: Several kinds of detectors have been used to signal the emergence of a component from the column. The most common types are *thermal conductivity cells* and *flame ionization detectors.* With the former, helium should be used as the carrier gas to get reasonable sensitivity since, with *thermal conductivity cells,* the resistance of a hot wire cooled by the gas stream emerging from the column is being measured, and there is a much greater difference in thermal conductivity between helium and most organic materials than there is between nitrogen and most organic substances. *Flame ionization* detectors measure the resistance between two electrodes on either side of a flame brought about by adding a small percentage of hydrogen to the carrier gas and burning it as it exits from the column. As combustible compounds emerge from the column, they also burn, producing many ions and decreasing the resistance between the electrodes. This type of detector is capable of much greater sensitivity, but it suffers from the disadvantage of destroying the sample; this drawback is minimized in most instruments by splitting the emerging gas stream and feeding only a small portion of the eluent into the detector.

Detectors are usually connected directly to recorders so that a graphical display in the form of peaks is presented as a permanent record.

The retention time of a substance is a characteristic of its structure and can be used to identify the material by comparison with a known sample. Since retention times vary as a function of experimental conditions, once the identity of a component is suspected, the unknown should be diluted with a small percentage of a known to see whether the height of the tentatively identified peak increases and to ascertain that no shoulder appears on the peak. Under these circumstances, the assumption of identity is sounder, particularly if the same results are obtained on several different columns. Even then, a conclusion of identity is not completely safe, and additional methods of comparison (Sec. 3.1) should be used.

Measurement of the area under a peak can lead to a quantitative value for the percentage of a component present in the mixture. It is important to use the *area* and not the peak height: Fig. 2-27 shows how the chromatogram of an equimolar mixture of three components might look. Note that while the area under each peak is about the same, the peak heights are very different, with the component with the shortest retention time giving the tallest, narrowest peak. Since the area under a peak may not be directly proportional to the relative percentage of that component in the original mixture (particularly with flame ionization detectors), it is important that known mixtures be investigated to establish the relationship between peak area and percentage composition. The area can be estimated by reading the integral curve (if the recorder has an integrator), by cutting out the peaks and weighing the paper, with a planimeter, by triangulation of the peaks, or (especially if the peaks are poorly resolved), with the aid of a "Curve Resolver."† With care, relative areas of well-resolved peaks can be estimated to within 1–2% routinely by any of these methods.

2.5.4 Collectors: Collection of emerging fractions is frequently complicated by aerosol formation, which can make it very difficult to condense the desired components. Condensation is facilitated by the presence of a large surface area; this is often accomplished by packing the receiver with glass wool. The compound can then be removed by washing the glass wool and vessel with a low boiling solvent that is easily removed by distillation. A simple collector that works well is illustrated in Fig. 2-29. The tube can be immersed in an ice or Dry Ice bath if the sample is particularly volatile, and most of the condensed liquid can be recovered *without* adding solvent by centrifuging the whole assembly.

For identifying small amounts of substances as they come off the column, the emerging vapors can be passed through a small cooled tube packed with dry potassium bromide. The amount of sample adsorbed by the salt is usually enough so that a pellet of the salt will give an excellent infrared spectrum. Another technique is to collect all the eluent and carrier gas in an evacuated bulb and then isolate the compound by distillation on a vacuum line.

† Instrument Products Division, E. I. du Pont de Nemours & Co., Wilmington, Del. 19898 ($10,000).

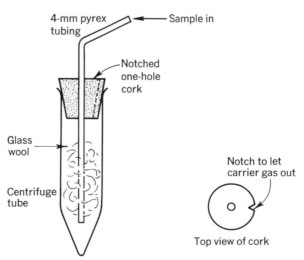

4-mm pyrex tubing

Sample in

Notched one-hole cork

Glass wool

Centrifuge tube

Notch to let carrier gas out

Top view of cork

Fig. 2-29 Collector for chromatography.

2.5.5 Gas Chromatographs: Commercial instruments are available for both analytical and preparative work, the former for $2,000–5,000 and the latter for $3,000–10,000. A workable setup can be assembled for much less, but considerable time is required. Most of the commercial instruments are available with temperature programming, a feature that can speed up compound separation times enormously since the temperature of the column can be gradually increased throughout a run, thereby drastically decreasing the difference in retention times between low and high boiling components. Some new preparative scale chromatographic instruments are reported to be capable of handling up to 100 ml of liquid in a single pass. Unfortunately, resolution decreases as column diameter increases. Some of the preparative instruments have small diameter columns in parallel to maintain high resolution; in such cases great care must be taken to pack the columns similarly so that a substance will have essentially the same retention time on each column. The more expensive preparative instruments have provision for automatic repetitive injection and collection.

2.6 LIQUID CHROMATOGRAPHY

Many of the principles that govern gas chromatography apply equally well to liquid chromatography (LC) with the obvious difference that the mobile phase is a liquid rather than a gas. Although liquid chromatography generally takes longer and requires more skill on the part of the experimentalist than gas chromatography does, it can be used with many compounds insufficiently volatile for GC, and it is somewhat more convenient for large scale separations. In the following variations of LC, the stationary phase is contained in a cylindrical column, as in GC: *adsorption chromatography, partition chromatography, ion-exchange chromatography,* and *gel permeation chromatography.* After these "column chromatography" or "elution chromatography" methods have been discussed, two LC methods in which the stationary phase is in the form of a thin sheet will be considered: *thin layer chromatography* (TLC) and *paper chromatography.* These thin sheet methods use much smaller samples and are much faster than the column methods; thus they are better suited for analytical separations. The column methods are generally used for preparative purposes.

2.6.1 Adsorption and Partition Chromatography: In chromatography employing liquids as eluents, the separation of a mixture is accomplished by passing a solution containing the sample through a tube partially filled with a stationary solid, which may or may not be coated with a tightly bound liquid (e.g., water on silica in silica gel). If a tightly bound liquid is present, the separation probably depends largely on the components of the sample being partitioned differently between the tightly bound liquid and the mobile liquid eluent, and the term *partition chromatography* or *liquid-liquid chromatography* is often applied. If a tightly bound liquid is not important and the separation requires that the sample components have different adsorption coefficients on the stationary solid, *adsorption chromatography* or *liquid-solid chromatography* is involved.

The adsorbents that have been used are numerous and varied in nature: They include charcoal, cellulose, Fuller's earth, silica gel, and alumina. The choice of adsorbent is based on the polarity and characteristics of the compounds to be separated, but in most instances an appropriate grade of activated alumina gives excellent results. In general, the polarity of the adsorbent should be significantly less than that of the substances to be separated.

In addition to variations of adsorption characteristics between materials like charcoal and alumina, wide variations are also possible within a class of adsorbents due to the presence of contaminants. The large diversity in properties of different brands and batches within a given brand is a major problem. If a difficult chromatographic separation is to be run more than once, a batch of adsorbent large enough for all the runs should be obtained so that it is necessary to work out the separation conditions only once. Much liquid chromatography is carried out on alumina, and the remainder of the discussion concerns its use, although the principles outlined apply to other substances as well.

Alumina for chromatography comes in three forms (basic, pH ∼ 10; neutral, pH ∼ 7; acidic, pH ∼ 4) whose activity can be varied quantitatively by the addition of measured amounts of water. The most active form of alumina ("activity I" or "Brockmann Number I") is too active for normal applications and is rarely used. Activity III alumina is best for most applications, and it should be used when alumina is first being tested as an adsorbent. If it is found that the compounds are adsorbed too tenaciously or an excessive amount of "tailing" of components occurs after the peaks, a less active grade of alumina is used. If the compounds elute rapidly and are insufficiently separated, a more active grade should be investigated. The activity can be checked with a special series of dyes.†

The efficiency of a separation is influenced by the activity of the adsorbent, the rate and polarity of the eluents, the skill of the operator in packing the column, and by the dimensions of the column. Table 2-4 contains a tabulation of weights of alumina and recommended column dimensions. As is apparent from this table, a 7.5:1 height:diameter ratio is considered optimum. Although the ratio of adsorbent to substrate required to give an acceptable separation varies, a 30:1 ratio is usually adequate.

When a column is to be prepared (Fig. 2-30), it is clamped in a vertical position and partially filled with a nonpolar solvent such as pentane. A small glass wool plug is pushed to the bottom of the tube with a glass rod and covered to a depth of about 1 cm with clean, coarse sand. A slurry of alumina in pentane is then added *slowly*, with vigorous tapping of the column to

---

† Obtainable from Alupharm Chemicals, Box 30628, New Orleans, La. 70130.

Table 2-4

LIQUID CHROMATOGRAPHY: WEIGHT OF ADSORBENT AND COLUMN
DIMENSIONS FOR VARIOUS SAMPLE SIZES

Sample, g	Adsorbent, g	Column inner diameter, mm	Adsorbent height in column, mm
0.01	0.3	3.5	25
0.03	1	5	40
0.1	3	7.5	55
0.3	10	11	80
1	30	16	120
3	100	24	170
10	300	35	250
30	1000	52	375
100	3000	75	550

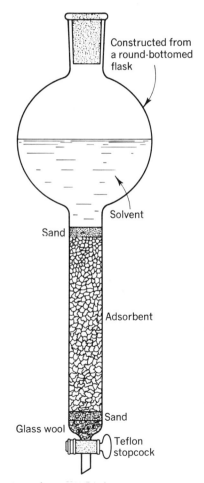

Fig. 2-30 Column chromatography apparatus.

dislodge air bubbles and encourage the alumina particles to settle. When all the alumina has been added, it is covered with a second layer of sand, and the solvent is drained from the column to a level just above that of the sand. A column should never be drained free of solvent since cracking and "channeling" are likely to result. The stopcock at the bottom of the column is preferably made of Teflon, but an ungreased glass stopcock can be used.

The sample is added to the column in a small volume of solvent (the least polar of the usual solvent sequence—given below—in which it will dissolve) and adsorbed by draining solvent from the bottom until the last bit of solvent above the alumina has disappeared. A small quantity of solvent is immediately added to the top of the column, and draining is carried out as before. Several such rinses suffice to transfer the mixture quantitatively to the adsorbent, and the column above the adsorbent is then filled with solvent and elution is begun. If the rate of solvent flow is excessively low, it can be accelerated by applying pressure to the top of the column by means of a syringe bulb or an air or nitrogen line.

A useful sequence of solvents is pentane, benzene, ether, ethyl acetate, methanol, acetic acid. A gradual change of solvents should be made by diluting the solvent with higher eluting power with the solvent directly preceding it and gradually increasing the enrichment on subsequent additions of solvent; in difficult separations, each drop is changed in composition through the use of a simple (and very inexpensive) technique called *gradient elution*. This involves adding the more polar solvent dropwise to the less polar solvent at the same rate that the latter is added to the column, and it is easily accomplished by placing a long-stemmed dropping or separatory funnel (without pressure-equalizing tube!) containing the more polar solvent above the solvent reservoir at the top of the column and opening wide the funnel stopcock [Fig. 2-31(a)]. If the column does not have a solvent reservoir at the top, the setup in Fig. 2-31(b) can be used. In either case, to promote mixing of the solutions, the stems of the dropping funnels should be *short* (as shown) if the upper solution is more dense and *long* if it is less dense.

It is rare that one has to go through the entire sequence of solvents during a particular chromatographic separation. To save time when a new separation is being attempted, a preliminary run can be made with rapid changing of solvents to determine when the desired product is eluted. It should be noted that alumina is appreciably soluble in methanol and acetic acid, and the white crystals of this substance formed on evaporation of solvent should not be mistaken for product.

Detection of products during this type of chromatography is generally far less convenient than in gas chromatography. If the desired substance is colored or fluoresces under the influence of ultraviolet light, its progress down the column is readily observed and the appropriate fractions are collected. The presence in the eluate of a substance of interest can often

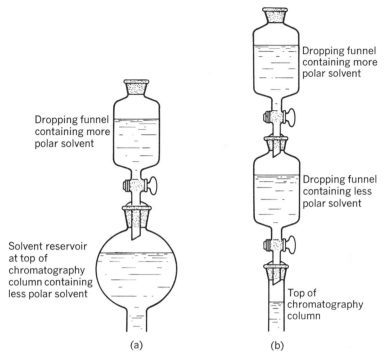

Dropping funnel containing more polar solvent

Dropping funnel containing more polar solvent

Dropping funnel containing less polar solvent

Solvent reservoir at top of chromatography column containing less polar solvent

Top of chromatography column

(a)   (b)

Fig. 2-31 Gradient elution setups.

be detected by a technique such as TLC (Sec. 2.6.5) or UV (Sec. 3.3.4). If not, the eluent is usually collected at regular volume intervals and the resulting fractions are evaporated to dryness and weighed. A plot of sample weight versus fraction number will give a graphical record of the progress of the chromatography. Automatic fraction collectors that change fractions at given time or volume intervals are available for $500–1000 (Fig. 2-32); refrigerated models which can be operated as low as 0° cost more ($1400–1800). Concentration of sample fractions is most easily accomplished with the aid of a rotary evaporator (Sec. 2.4.6).

It cannot be overemphasized that care and patience on the part of the investigator is a particularly important requirement for a successful separation through adsorption or partition chromatography.

2.6.2 Ion-exchange Chromatography:† In this technique, the stationary phase is a resin containing functional groups that interact with ions and substances that are easily ionized. It is used extensively for changing cations or anions, for deionizing solutions, and for separating similar polar sub-

† F. Helfferich, *Ion Exchange,* New York, McGraw-Hill Book Co., 1962; C. Calmon and T. R. Kressman, *Ion Exchangers in Organic and Biochemistry,* New York, Interscience Publishers, 1957.

Fig. 2-32 Automatic fraction collector.

stances such as the amino acids obtained by hydrolyzing proteins.† Table 2-5 lists commercially available resins of various types. The strongly acidic resins are ordinarily used for cation exchange and the strongly basic resins for anion exchange, provided there are no substances involved which are unstable at the pH extremes encountered with these resins.

In preparing an ion-exchange column, the resin should first be fully hydrated with deionized water in a beaker because if it is put on the column dry, the swelling pressure when water is added may burst the column. The

† "Amino acid analyzers" in which the hydrolysis of a protein and the separation of the resulting amino acid mixture by ion-exchange chromatography are automated can be purchased for $12,000–25,000.

column is filled with the resin-water slurry until the resin half fills the column. A water line is then attached to the bottom of the column, and it is "backwashed" (Fig. 2-33) by very slowly running deionized water through this line until any air pockets have been removed and all resin particles are free-flowing. This step sorts the resin particles according to size, with the smallest at the top; in doing so, it increases the efficiency and flow properties of the column. The resin is then allowed to settle. If the resin is in the desired ionic form, the column is ready for use.

The sample is applied to the top of the column in much the same way as in the previous section, except that the solvent is generally water or a buffered aqueous solution. Elution is brought about by passing more solvent through the column.

The process of rapidly changing the ionic form of a column, termed *regeneration,* is applied to new resin when it has been obtained in an undesired ionic form and to resin which has been *exhausted* in an ion-exchange experiment. After a preliminary backwashing, the regenerant solution is added slowly to the column. The amount of regenerant used is 1.5–5 times the theoretical amount required, and its concentration is about 1 M or 4–10% by weight. Sometimes the effluent is tested to make sure sufficient regenerant solution has been added. Cationic exchange resins are converted to $H^{\oplus}$ with hydrochloric or sulfuric acid and to $Na^{\oplus}$ with sodium chloride (strongly acidic) or sodium hydroxide (intermediate or weakly acidic). Strongly basic anion exchange resins are converted to $OH^{\ominus}$ with sodium hydroxide and to $Cl^{\ominus}$ with hydrochloric acid or sodium chloride;

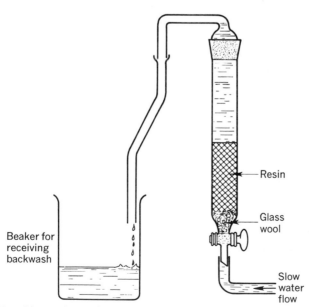

Fig. 2-33 Backwashing an ion-exchange column.

Table 2-5

ION-EXCHANGE RESINS

Type and exchange group	(1) Bio-Rad Labs	(2) Dow Chemical Company "Dowex"	(3) Diamond-Shamrock Corp. "Duolite"	(4) Rohm & Haas Co. "Amberlite"	(5) Permutit Company (England)	(5) Permutit Company (U.S.A.)	(6) Nalco Chemical Co. "Nalcite"
Cation exchange resins							
Strongly acidic, phenolic type $R\text{-}CH_2SO_3^-H^+$	Bio-Rex 40		C-3		Zeocarb 215		
Strongly acidic, polystyrene type $\phi\text{-}SO_3^-H^+$	AG 50W-X1 AG 50W-X2 AG 50W-X4 AG 50W-X5 AG 50W-X8 AG 50W-X10 AG 50W-X12 AG 50W-X16	50-X1 50-X2 50-X4 50-X5 50-X8 50-X10 50-X12 60-16	C-25D C-20 C-20X10 C-20X12	IR-112 IR-120 IR-122 IR-124	Zeocarb 225 (X4) Zeocarb 225	Permutit Q Q-100 Q-110 Q-130	HCR HGR HDR
Intermediate acid, polystyrene type $\phi\text{-}PO_3^=(Na^+)_2$	Bio-Rex 63		ES-63				X-219

	Bio-Rex 70		CC-3	IRC-84 IRC-50	Zeocarb 226	Q-210	
Weakly acidic, acrylic type $R\text{-}COO^-Na^+$	Bio-Rex 70					Q-210	
Weakly acidic chelating resin, polystyrene type $\phi\text{-}CH_2N\big\langle{}^{CH_2COO^-H^+}_{CH_2COO^-H^+}$	Chelex 100	A-1					
Anion exchange resins							
Strongly basic, polystyrene type $\phi\text{-}CH_2N^+(CH_3)_3Cl^-$	AG 1-X1 AG 1-X2 AG 1-X4	1-X1 1-X2 1-X4	A-101D	IRA-401	DeAcidite FF (lightly cross-linked)	S-100	
	AG 1-X8	1-X8			DeAcidite FF		SBR
	AG 1-X10 AG 21K AG 2-X4	1-X10 21K 2-4	A-102D	IRA-400			
$\phi\text{-}CH_2N^+(CH_3)_2(C_2H_4OH)\,Cl^-$	AG 2-X8 AG 2-X10 Bio-Rex 9	2-X8		IRA-410		S-200	SBR-P
NH^+Cl^- (pyridinium)						S-180	SAR

Table 2-5

ION-EXCHANGE RESINS (CONT.)

Type and exchange group	(1) Bio-Rad Labs	(2) Dow Chemical Company "Dowex"	(3) Diamond-Shamrock Corp. "Duolite"	(4) Rohm & Haas Co. "Amberlite"	(5) Permutit Company (England)	Permutit Company (U.S.A.)	(6) Nalco Chemical Co. "Nalcite"
Intermediate base, epoxypolyamine $R\text{-}N^+(CH_3)_3Cl^-$ and $R\text{-}N^+(CH_3)_2(C_2H_4OH)\,Cl^-$	Bio-Rex 5		A-30 A-30B		F	S-310 S-380	
Weakly basic, polystyrene, or phenolic polyamine $R\text{-}N^+HR_2Cl^-$ $R\text{-}N^+HR_2Cl^-$	AG 3-X4	3-X4	A-2 A-6 A-7	IR-45 IR-4B IRA-68	G	S-300 S-350	WBR
Ion retardation (zwitterion) resin							
Strongly basic anion plus weakly acidic cation exchange resin $\phi\text{-}N^+(CH_3)_3Cl^-$ and $R\text{-}CH_2COO^-H^+$	AG 11A 8						
Redox resin							
$-NR_2Cu^0$			S-10				

Mixed bed resins					
ϕ-SO$_3$$^-H^+$ and ϕ-CH$_2$N$^+$(CH$_3$)$_3$OH$^-$	AG 501-X8	GPM-331G	MB-1	Bio-Demineralit	M-100
ϕ-SO$_3$$^-H^+$ and ϕ-CH$_2$N$^+$(CH$_3$)$_3$OH$^-$ indicator dye	AG 501-X8 (D)			Indicator Bio-Demineralit	M-103
ϕ-SO$_3$$^-H^+$ and ϕ-CH$_2$N$^+$(CH$_3$)$_2$ (C$_2$H$_4$OH)OH$^-$		GPM-331A	MB-3		

(1) Bio-Rad Laboratories, 32nd and Griffin Ave., Richmond, Calif. 94804.
(2) Dow Chemical Co., 2030 Abbott Road Center, Midland, Mich. 48640.
(3) Diamond-Shamrock Corp., 300 Union Commerce Bldg., Cleveland, Ohio 44115.
(4) Rohm & Haas Co., Philadelphia, Pa. 19105; distributed by Industrial Chemicals Division, Mallinkrodt Chemical Works, St. Louis, Mo. 63147.
(5) Permutit Co., E. 49 Midland Ave., Paramus, N.J. 07652.
(6) Nalco Chemical Co., 180 N. Michigan Ave., Chicago, Ill. 60601.

weakly basic anion resins require hydrochloric acid for conversion to the $Cl^{\ominus}$ form, and they are converted to the free base form by sodium or ammonium hydroxide or sodium carbonate. The column is rinsed with about 10 times the "bed volume" (volume occupied by the resin beads and the liquid among them) of deionized water to remove excess regenerant.

2.6.3 Gel Permeation Chromatography: This technique, also called *gel filtration,* is used for separations involving macromolecules. As in ion-exchange chromatography, the stationary phase consists of a cross-linked resin, but in this case the resin beads contain pores large enough to hold certain size molecules, and the separation is based almost entirely on molecular size. Molecules small enough to fit into the pores are held back, whereas molecules too large pass directly through. Compounds that fit in the pores are then eluted in order of decreasing molecular size. The sharpest separations are between substances smaller than the "exclusion limit" for the resin and those that are not; resins are available with exclusion limits ranging from molecular weight 1,000–300,000.

The column is prepared and used as in ion-exchange chromatography. It is important that the resin be soaked with solvent for 2–48 hr before placing it in the column.

Gels commonly used with polar substances (and aqueous solutions) are Sephadex,† Bio-Gel,‡ and Bio-Beads‡. The last of these are available in a form usable with nonpolar substances and solvents.

This technique has been used extensively by polymer chemists to characterize the molecular weight distribution of a polymer preparation and by biochemists to desalt or separate enzymes, nucleic acids, and polysaccharides. When the procedure is applicable, it can be used very effectively on a preparative scale.

2.6.4 Paper Chromatography: In this technique, used primarily for analysis, the stationary phase consists of a strip of high quality filter paper, which is occasionally modified by impregnation with a material such as alumina or an ion-exchange polymer. A small quantity of the mixture is spotted on the paper with a glass capillary, and the paper is placed in a solvent chamber (Fig. 2-34). The solvent moves through the paper by capillary action, moving components of the mixture along with it, hopefully at different rates. These rates are expressed as R_f values, defined as the ratios of the distances which the samples migrate as compared to the distance from the starting point to the solvent front.

For *ascending* paper chromatography [Fig. 2-34(a)], a strip of paper is ruled near the bottom edge and drops of the sample and, if possible, known standards are spotted along the line. The paper is then suspended in an

† Pharmacia Fine Chemicals, Inc., 800 Centennial Ave., Piscataway, N.J. 08854.
‡ Bio-Rad Laboratories, 32nd & Griffin Ave., Richmond, Calif. 94804.

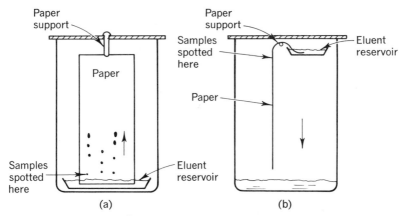

Fig. 2-34 Paper chromatography.

enclosed airtight tank with the bottom edge immersed in a tray of solvent. The eluent moves up through the paper and when the solvent front nears the top of the paper, the paper is removed from the jar and the edge of the solvent front marked. The paper can then be sprayed with a reagent that develops a spot where each component of the mixture resides, or the positions can be detected by other means such as fluorescence.

Components can be identified from R_f values and, in favorable cases, by their spectral properties. This can be accomplished by cutting out the section of paper containing a component, soaking the paper in a solvent, and then recording the spectrum of the component in the solvent.

In a variation called *descending* paper chromatography [Fig. 2-34(b)], the solvent tray is placed at the top of the jar and the solvent is allowed to move *down* the strip of paper. The flow of solvent is much more rapid in this case since it is aided by gravity. By having the paper rise out of the tray slightly before descending, a too rapid siphoning action is avoided.

A simple extension of this technique that has increased its power as an analytical method is known as *two-dimensional paper chromatography*. In this approach, the paper is spotted at one corner and the sample chromatographed in the usual manner. When the solvent front reaches the top of the paper, the paper is removed from the chamber, dried, and then placed in a second chamber containing a different solvent. This time, however, the paper is mounted at right angles to its original direction. In this way, two R_f values from solvents varying in eluting power can be obtained to aid in the identification of unknowns and in the separation of complex mixtures such as the amino acids present in protein hydrolysate.

Separations worked out on paper chromatography can be scaled up for preparative purposes by employing a column packed with powdered filter paper and the same eluents (Sec. 2.6.1).

2.6.5 Thin Layer Chromatography:† This name (abbreviated TLC) is applied to the "thin layer" versions of the column techniques described in Secs. 2.6.1 to 2.6.3, other than paper chromatography. The stationary phases (except cellulose) mentioned in these sections lack the strength of paper and require the support provided by a glass plate or polyester film.

The commonest stationary phases for TLC are silica gel and alumina, the latter usually mixed with calcium sulfate (plaster) as a binder. Ion-exchange and gel filtration beads are available in finer mesh sizes for TLC.

Rolls and sheets of thin layers on polyester films can be purchased ready to use,‡ or glass plates can be coated in the laboratory. For small scale experiments, a microscope slide can be coated by simply dipping it into a slurry of stationary phase (e.g., 30% silica gel in chloroform). To coat large plates uniformly on one side, a spreader—either commercial or constructed by putting two collars of adhesive tape or rubber tubing around a glass rod—is used (Fig. 2-35). The spreader is drawn or rolled along in one motion to produce a uniform layer of stationary phase about 0.3 mm thick. The coated plate is in most cases air dried and then oven dried before use.

Samples are spotted on the plate, which is developed in a solvent chamber as described for paper chromatography. The solvents used for TLC are the same ones used in column chromatography on the same stationary phase.

To permit visualization of the spots, the plate is often sprayed with something; e.g., silica gel plates show spots for most organic substances when the plates are sprayed with a methanol solution of sulfuric acid and then baked in an oven. A permanent record can be made by photographing the plates in color before the colors fade. An alternative method for detecting spots consists of mixing about 1% of a fluorescent or phosphorescent substance (sold by companies which handle TLC accessories) with the adsorbent

† A good text on this subject is E. Stahl and Associates, *Thin-Layer Chromatography—A Laboratory Handbook,* New York, Academic Press, Inc., 1965. An up-to-date bibliography on TLC and a catalog of TLC accessories can be obtained by writing the Kensington Scientific Corporation, Box 531, Berkeley, Calif. 94701.

‡ Eastman Organic Chemicals Department, Eastman Kodak Co., Rochester, N.Y. 14603; Industrial Chemicals Division, Mallinkrodt Chemical Works, St. Louis, Mo. 63147.

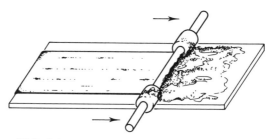

Fig. 2-35 Coating a TLC plate.

before it is applied to the plate and, after development, observing it cautiously under ultraviolet light. Dark spots are observed where the components of the mixture are, due to quenching of the fluorescence.

As an analytical tool, TLC is much less quantitative than gas chromatography, but it has the advantage that a large number of samples can be run simultaneously and thus in less time.

Separations worked out by TLC are readily scaled up by "streaking" a larger amount of the mixture on a large TLC plate or by going to columns with the same adsorbent and eluents.

3
Structure
Determination
Techniques

Almost all organic chemists spend a significant fraction of their time characterizing organic compounds. The complexity of characterization problems varies greatly, from simple cases like confirming that the lithium aluminum hydride reduction product of a simple aliphatic ketone is the expected secondary alcohol (an available solid of known structure) to such difficult cases as the elucidation of the stereoformula of a new natural product containing hundreds of atoms and having thousands of stereoisomers.

A wide variety of methods which may be applied toward the solution of such problems are available, and the chemist must use his ingenuity in the selection of the best approach, i.e., that one which most efficiently leads to the structure with a minimum of ambiguity. In the former example above, the best approach would consist of establishing the identity of the resulting alcohol with an authentic sample, probably by showing the superimposability of their infrared and NMR spectra and the failure of a mixture to show a decreased melting point. In the latter case, spectra and elemental analyses would be obtained, and if it appeared unlikely that the structure could be established readily by relating the compound chemically to a compound of known structure, a single crystal of a heavy atom derivative of the natural product would be prepared and X-ray diffraction methods applied. Since all the available methods are subject to ambiguities which leave some room for doubt regarding the structure, two or more independent lines of argument are often sought. For example, structures based on degradative evidence are often checked by synthesis.

With the variety of methods available, it is impractical to outline a scheme which applies for all unknowns. Rather, the most useful and widely applied methods are discussed, and some of the other methods are mentioned briefly.† Although structures can be deduced for components of mixtures without isolation of the pure substances, this is usually successfully

† For more extensive treatment of almost every method described in this chapter, see D. J. Pasto and C. R. Johnson, *Organic Structure Determination,* Englewood Cliffs, N.J., Prentice-Hall, Inc., 1969.

avoided by the application of the powerful separation methods now available (Chapter 2), and it is assumed below that the unknown is a pure substance.

One of the major advances in organic chemistry in the last decade has been the shift in emphasis from chemical to physical methods for structure determination, made possible by the development of many versatile physical tools. The major advantages of physical methods are ease of application and freedom from the ambiguity due to the possibility of chemical rearrangement. Chemical methods are still of great value, but they are usually not employed until after the readily accessible physical methods have been exhausted, and then only when the physical methods do not give a sufficiently conclusive answer.

3.1 IDENTITY WITH AN AUTHENTIC SAMPLE

If the unknown has been described in the literature and is well characterized, to establish the structure of the unknown it is necessary only to demonstrate that the two substances have the same structure. If the unknown has not been previously described, its structure determination may well involve showing that one or more of its degradation products are identical with known compounds. Thus, it is valuable to assess the ways of showing that two samples consist of the same substance.

All the chemical and physical properties of the two samples must be the same, but certain ones make much better "fingerprints" than others. Although X-ray powder photographs (see below) are the only infallible fingerprints, they are not often employed by organic chemists, who instead usually use a *combination* of less reliable but more familiar techniques. None of these techniques requires more than a few milligrams of unknown, and some of them can easily be applied with microgram samples. The equipment requirements for the spectral methods are discussed later. Many of these methods require that an authentic sample of the known be on hand for comparison purposes. If the substance is not commercially available, an authentic sample can often be obtained by writing to a chemist who has recently prepared it, or it can be synthesized. If an authentic sample is not obtained, the best that can be done is to show that the unknown has all the properties mentioned in the literature for the compound (e.g., same melting point and/or boiling point, same melting points for derivatives;† spectra—especially infrared—have been published for tens of thousands of compounds).‡

Infrared, NMR, and mass spectra are usually reliable as fingerprints. Mixed melting point determinations with solid unknowns or with solid derivatives of liquid unknowns rarely give misleading results. Identity should

---

† Derivatives for over 8000 compounds are listed in *Handbook of Tables for Organic Compound Identification,* Cleveland, Ohio, Chemical Rubber Co.

‡ *The Sadtler Standard Spectra,* Philadelphia, Pa., Sadtler Research Laboratories, Inc.

not be concluded on the basis of identical retention time on one chromatography column, but if two substances have the same retention times on several different columns, it is very likely that they have the same structure. If the compound has a reasonably rare property such as an unusual biological activity or color test, this should be applied. If the compound is optically active, its optical rotation (or better, optical rotatory dispersion curve) should be measured; none of the other methods (except biological activity and mixed melting points) will distinguish between enantiomers.

As noted above, the most powerful comparison technique is X-ray powder photography:† No two different compounds (other than mirror images) have ever been found to give the same powder pattern. This method is limited to crystalline substances and, since few X-ray powder photographs have been published for organic substances, to cases in which the authentic sample is on hand. Only a few milligrams is necessary. It is important that the samples be recrystallized in the same manner since many organic compounds crystallize in more than one form, depending on conditions; if there is a chance that two different crystalline forms of the same substance are being compared, a finding of nonidentity will be meaningless. The photographs can easily be taken in a few hours. The necessary equipment, an X-ray generator ($4000) and a powder camera ($1000), is available in many chemistry, physics, and geology laboratories.

With the comparison methods above, it is generally easy to establish the identity of an unknown with a previously described compound, especially if an authentic sample of the known is on hand. The major problem is usually recognizing with which of the millions of compounds of known structure to make the comparison; this is often not apparent until some of the methods described in the next sections have been used to reveal something of the structure.

3.2 MOLECULAR FORMULA

With the growing availability of mass spectrometers of sufficiently high range and resolving power for most organic unknowns, mass spectrometry has become the most useful tool for determining molecular formulas. It is discussed at length before the other methods.

Mass spectra have long been used (a) as fingerprints for organic compounds (previous section) and (b) for the determination of accurate molecular weights (well within one mass unit). More recently, they have been exploited (c) for the determination of exact molecular formulas. The emphasis in this section is on applications (b) and (c).

---

† For a discussion of how to measure a powder pattern, as well as a catalog of several thousand patterns of known compounds, see L. J. E. Hofer, W. C. Peebles, and E. H. Bean, *X-ray Powder Diffraction Patterns of Solid Hydrocarbons, Derivatives of Hydrocarbons, Phenols, and Organic Bases, Bulletin 613, Bureau of Mines,* Washington, U.S. Govt. Printing Office, 1963.

The chief difficulty with mass spectrometric molecular weights is that a small percentage are, for several different reasons, off by one or more mass units, thus introducing an element of doubt in molecular weights obtained in this way. To appreciate the techniques used to minimize these sources of error, it is necessary to know something of the changes occurring in a sample between its introduction into a mass spectrometer and the appearance of peaks from it in the spectrum.

The sample is vaporized and exposed to a beam of electrons, producing neutral and positively and negatively charged species. The negatively charged species can be observed if desired, but mass spectrometers are usually set to scan for cations. In most mass spectrometers, the positive species are accelerated in a fixed electric field and the resulting beam is bent by a variable magnetic field toward a fixed ion detector connected to a recorder. The magnetic field serves to separate the species according to their ratio of mass to charge (a very small fraction of the cations have more than one unit of positive charge) since cations with a higher mass to charge ratio are moving much more slowly and are thus more influenced by the magnetic field. The strength of the magnetic field is gradually decreased so that the desired range of mass to charge ratios is scanned from low to high values. The mass spectrum (Fig. 3-1) is a plot of relative abundance of positively charged species against their mass to charge ratios. To conserve space, the data are generally plotted as shown in this figure (the most intense peak in the spectrum is called the "base peak").†

† Spectrometers whose direct output is nearly in this desirable form can be purchased from Varian Associates, Palo Alto, Calif.

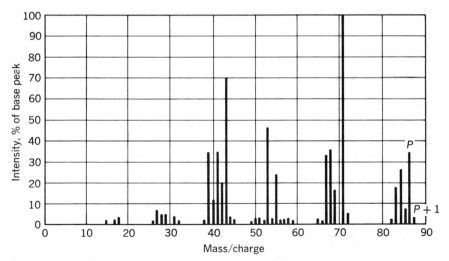

Fig. 3-1 Mass spectrum of γ, γ-dimethyl alcohol.

One of the commonest occurrences which generates a cation is the collision of an electron with a molecule to yield two electrons and a cation radical, called the *molecular ion* since it has essentially the same mass as the original molecule. Fortunately, the peak due to the molecular ion (the "parent peak," P) is usually easily identified in the spectrum, thus yielding the molecular weight of the sample. The molecular ion peak is generally the largest peak in the group of peaks (differing by one mass unit) at *highest mass* since many fragmentations to give cations of lower mass occur, but few collisions result in the formation of cations of higher mass than the molecular ion. For organic compounds, there are always small "isotope peaks," (e.g., $P + 1$ and $P + 2$ peaks) at higher mass than the parent peak, due to the presence of molecular ions containing one or more of the heavy isotopes of the elements present, e.g. ^{13}C (natural abundance, 1.12% of ^{12}C) and ^{2}H (natural abundance, 0.016% of ^{1}H). As described below, these peaks can be useful in the determination of molecular formulas.

A common situation in which the identification of the parent peak is difficult occurs when the $P + 1$ peak is abnormally large (sometimes much larger than the parent peak). This is due to collisons of molecular ions with proton-containing species in such a way that a proton is abstracted; this is characteristic of many compounds containing hetero atoms, such as ethers and amines. Fortunately, this sort of collision is the only common type which gives peaks at higher mass to charge ratio than the parent peak. This situation can easily be detected by increasing the sample pressure (or, more easily and with the same effect, decreasing the *repeller voltage,* which gives the ions an initial push toward the accelerating voltage), causing more collisions and making the $P + 1$ peak larger with respect to most of the other peaks.

A second test which is easily applied serves to distinguish molecular ion peaks from fragmentation peaks, the most troublesome of which is usually the $P - 1$ peak because of its proximity to the high end of the spectrum. This involves *lowering the voltage of the ionizing beam of electrons to just above the "appearance potential"* (the minimum voltage at which molecular ions are produced), and it greatly increases the size of the molecular ion peak relative to fragmentation peaks.

The coincidence that organic compounds containing the usual elements have even-numbered masses except when an odd number of nitrogens is present is often of considerable aid in the selection of the molecular ion peak.

Even when these two tests have been applied, the possibility exists that the molecular ion is so unstable that it does not appear in the spectrum. This is especially true for tertiary alcohols. In these cases, it is sometimes possible to determine the exact molecular weight by studying the fragmentation pattern. Tertiary alcohols, for example, usually give strong peaks at $P - 18$ (loss of water from the molecular ion).

There are two ways of deriving exact molecular formulas from mass spectra, each successful only when the parent peak has been properly identified as described above. The first takes advantage of the slight differences in masses which exist for various molecular formulas. For example, $C_{10}H_{16}O_4$, $C_{12}H_{24}O_2$, and $C_{13}H_{16}N_2$ are three of the possible molecular formulas for a compound with a parent peak at 200. If the position of the parent peak can be determined with sufficient accuracy, however, they can be distinguished, since more exact values are 200.105, 200.178, and 200.131, respectively. "High-resolution mass spectrometers" capable of distinguishing these species cost at least $60,000, twice the minimum for usable mass spectrometers. If such an instrument is accessible and if the compound gives a recognizable molecular ion peak, the molecular formula can be determined by this method.† The possible confusion of the parent peak with $P + 1$ and $P - 1$ does not arise with high resolution, since the $P + 1$ and $P - 1$ molecular formulas will not correspond to stable species.

Even with a medium resolution mass spectrometer, it is often possible to obtain the molecular formula by examining the sizes of the isotope peaks at $P + 1$, $P + 2$, etc., mentioned above. The natural abundances of various isotopes are known accurately, and the sizes of the isotope peaks relative to the parent peak can be easily calculated for any molecular formula. The results of such calculations for the most common molecular formulas have been published.† The number and kind of many common hetero atoms such as the halogens and sulfur are readily revealed by the pattern of these isotope peaks. As an example of the use of isotope peaks, the three molecular formulas with mass about 200 in the example above could be distinguished in this way. For the first, the $P + 1$ peak is 11.22% of the height of the parent peak and the $P + 2$ is 1.37% of P; for the second, these values are 13.43% and 1.23%; for the third, 15.07% and 1.06%. The differences between the percentages for each of these molecular formulas are considerably greater than experimental error. It is necessary to be careful of interference from $P + 1$ peaks due to hydrogen abstraction as mentioned above and from the isotope peaks on $P - 1$ peaks when these are large. These interferences can be minimized by working at low pressures and low ionization potentials.

A major limitation of mass spectrometry is that the sample must be volatilized to provide an appreciable vapor pressure in the inlet system. Most

† Tables of masses and sizes of $P + 1$ and $P + 2$ peaks for all organic species containing only C, H, O, and N in the mass range 12–500 have been published (J. H. Beynon and A. E. Williams, *Mass and Abundance Tables for Use in Mass Spectrometry*, New York, Elsevier Publishing Co., 1963); exact masses for compounds containing no more than 30 carbons, 2 halogens, 2 phosphorus atoms, 5 nitrogens, 4 sulfurs, 9 oxygens, and with a limit of two kinds of heteroatoms are listed by M. J. S. Dewar and R. Jones, *Computer Compilation of Molecular Weights and Percentage Compositions for Organic Compounds*, Elmsford, N.Y., Pergamon Press Inc., 1969. For compounds not covered in these books, calculations can be performed as described by J. Lederberg, *Computation of Molecular Formulas for Mass Spectrometry*, San Francisco, Calif. Holden-Day, Inc., 1964.

instruments have heatable inlet systems, which extend considerably the range of compounds which can be examined; however, the instability of most organic compounds at high temperatures causes difficulty. The best arrangement for relatively nonvolatile substances is to vaporize the sample by warming it in the immediate vicinity of the ionizing beam, and some instruments have provision for doing this. Alternatively, it may be possible to convert a nonvolatile substance to a more volatile derivative, e.g., a trimethylsilyl derivative from an alcohol (Sec. 2.5).

In summary, with an unknown of sufficient volatility, mass spectrometry is one of the first techniques that should be applied if a mass spectrometer is accessible. In most cases, a fraction of a milligram of material provides a wealth of information: the exact molecular weight, the molecular formula, an almost unambiguous fingerprint, and (as noted later) further structural information from the cracking pattern.

The other methods for obtaining molecular weights and empirical formulas are usually applied by experienced microanalysts in service laboratories, and it is only necessary for the organic chemist to provide the microanalyst with a highly pure sample of the unknown and to deduce the correct molecular formula from the raw data provided by the microanalyst. Each determination requires a few milligrams of material and each costs $5–10 in commercial laboratories. Quantitative elemental analyses are available for all the elements commonly encountered in organic molecules, and the results allow the calculation of the empirical formula. Considerable calculation time can be saved for compounds containing only hydrogen, oxygen, sulfur, and 50 or fewer carbons through the use of anticomposition tables,† which list all the possibilities in order of increasing percentage of carbon. Also, composition tables are available‡ which make it unnecessary to calculate percentages for most compounds of C, H, O, N, S, P, and the halogens. If the molecule is large (e.g., $C_{23}H_{38}O_2$), more than one molecular formula, usually differing by two hydrogens, may lie within the limits of experimental error (in this example, $C_{23}H_{36}O_2$ and $C_{23}H_{40}O_2$; $C_{23}H_{37}O_2$ and $C_{23}H_{39}O_2$ need not be considered unless it is suspected that the compound is a radical). Such elemental analyses formerly were required by journal editors for all new compounds, but it should now be possible to substitute a mass spectrometrically determined molecular formula. Molecular weights are determined commercially by osmometry, by the Rast method, ebullioscopically, cryoscopically, and by isothermal distillation; for polymer molecules, other techniques are used.§ When the empirical formula and molecular weight are known, the determination of the molecular formula is completed.

† H. H. Hatt, T. Pearcey, and A. Z. Szumer, *Anti-Composition Tables for Carbon Compounds,* New York, Cambridge University Press, 1955.

‡ For CHONS compounds with up to 40 carbons, see G. H. Stout, *Composition Tables,* New York, W. A. Benjamin, Inc., 1963; for compounds of these elements plus phosphorus and the halogens, with up to 30 carbons, see M. J. S. Dewar and R. Jones, loc. cit.

§ P. J. Flory, *Principles of Polymer Chemistry,* Ithaca, N.Y., Cornell University Press, 1953.

3.3 MAJOR RAPID METHODS FOR STRUCTURAL FORMULA, STEREOFORMULA, AND CONFORMATION

For the organic chemist, a structure determination has just begun with the determination of the molecular formula (unless the molecular formula is so simple that there is only one isomer!) since the structural formula must be derived and any stereochemical questions must be answered. Even beyond the stereoformula, information can be sought regarding the preferred conformation(s) of the molecule, the bond lengths, and the bond angles. The techniques most commonly applied to gain further structural information are described before those that are less revealing or more time consuming. The first three are spectral techniques which can be profitably applied to almost any unknown; the remainder are less widely applicable.

3.3.1 NMR Spectrometry: If a nuclear magnetic resonance (NMR) spectrometer is available ($20,000–160,000), the NMR spectrum should be obtained since, of all the spectral methods, this one gives, in general, the most useful information. Infrared and mass spectra *contain* considerable structural information but much of it is virtually impossible to extract.

The NMR spectrum is usually obtained on a 1–30% solution (liquids may be run neat, but a larger sample is required and, due to intermolecular effects, the interpretation of the spectrum is slightly more difficult) in an aprotic solvent (usually deuteriochloroform, carbon tetrachloride, or deuterium oxide) containing a few percent of a reference compound (tetramethylsilane, abbreviated TMS; or, when deuterium oxide is used, the much more soluble sodium 3-(trimethylsilyl)-1-propanesulfonate).† Substances that are insufficiently soluble at room temperature are often soluble enough to give good spectra at higher temperatures; variable temperature probes usable between $-60°$ and $+200°$ are available for about $1500 for most spectrometers. About 30 mg of sample is required for a high quality spectrum, but somewhat inferior spectra yielding as much structural information can be obtained with as little as 1 mg through the use of microcells,‡ which employ a much smaller volume of solution carefully placed in the most sensitive region of the detector, and devices such as the "CAT" (computer of average transients, about $11,000), which scan rapidly over the spectrum many times and record the average, thereby increasing the signal to noise ratio by a factor equal to the square root of the number of traces (e.g., 100 traces increase the ratio by a factor of 10). For the most widely distributed instrument, the Varian A-60, about 10 min is required for the exacting task of tuning the instrument to the maximum performance condition, after which dozens of good spectra, each requiring about 10 min, can usually be obtained without retuning.§ A special glass tube containing the sample

† Available from Peninsular Chemresearch, Inc.. Box 14318, Gainesville, Fla. 32601.
‡ Available for about $10 from Kontes Glass Co., Vineland, N.J. 08360.
§ The Varian T-60, which has a permanent magnet rather than an electromagnet, is easier to tune and retains the tune much longer.

is spun between the poles of a very powerful magnet while radio waves are sent through it. A radio-frequency detector is used to tell how much radiation the sample is absorbing. Either the radio frequency or the magnetic field strength is varied slightly in the region of absorption while the other is held constant, and a recorder gives the NMR spectrum (Fig. 3-2) as a plot of intensity of absorption versus radio frequency or magnetic field strength.

The mechanism of absorption is complex. Certain kinds of nuclei, the most important for the organic chemist being protons, absorb. The absorptions of NMR-active nuclei other than protons occur in widely differing regions of the spectrum and, except for F^{19}, their detection requires additional expensive equipment ($6000 per isotope); they are not considered further.†

Each of the protons in the sample molecule gives essentially the same area of absorption in the spectrum, greatly facilitating the interpretation of NMR spectra.

The location of the peak(s) on the frequency scale (its "chemical shift") depends on the magnetic environment of the absorbing proton. Extensive tabulations of empirically determined chemical shifts for various types of protons are available, but it is important for the organic chemist to learn some of the commoner chemical shift values given in Tables 3-1, 3-2, and

† For a discussion of other isotopes, see J. W. Emsley, J. Feeney, and L. H. Sutcliffe, *High Resolution Nuclear Magnetic Resonance Spectroscopy,* Vols. 1 and 2, New York, Pergamon Press, 1965.

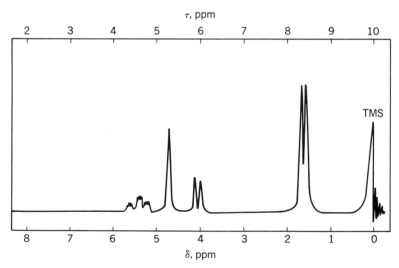

Fig. 3-2 NMR spectrum of γ, γ-dimethylallyl alcohol, 10% in DCCl$_3$ at 60 MHz.

3-3.† There has unfortunately not been general agreement on a frequency scale, and the situation has been complicated by the gradual shift from expressing frequencies in cycles per second (cps) to Hertz (Hz). The two scales used most often are the δ scale (δ = frequency of absorption in Hertz from tetramethylsilane/frequency of absorption in megaHertz) and the τ scale ($\tau = 10 - \delta$). These scales, both with the units "parts per million" (ppm), were defined to give convenient size numbers (most protons in organic molecules absorb between 0 and 10) which would be independent of the radio frequency used (usually 60 or 100 MHz). The latter scale has the advantage that larger numbers refer to protons absorbing farther "upfield," i.e., at higher frequencies and higher magnetic field strengths.

Table 3-2 shows that the replacement of a proton on a tetrahedral carbon atom by almost any grouping (silicon being an exception) causes a downfield shift in the absorption of any protons remaining on the tetrahedral carbon atom. In the table, the effect of one α-substituent is shown; if two or three α-substituents are present, the downfield shift is a little less than that calculated by summing the two or three shifts indicated in Table 3-2. If the substituent is β instead of α, it generally causes a shift about one-third as great as that indicated in Table 3-2, and in the same direction. The chemical shifts of aromatic protons (Table 3-3) are often useful in locating substituents on aromatic rings, particularly when the spin-spin splitting patterns are taken into account.

Spin-spin splitting (coupling) is a phenomenon observed in NMR spectra which in general considerably increases their value in structure determination. Coupling occurs only between NMR-active nuclei, and since it occurs through covalent bonds, it is exclusively an intramolecular phenomenon. Coupling constants ("J"), reported in Hertz (Hz) or cycles per second (cps) rather than ppm since they are independent of the magnetic field strength, fall off in magnitude rapidly as the number of covalent bonds connecting the coupled nuclei increases. For protons attached to tetrahedral carbons, when the separation is two covalent bonds ("geminal" protons), J ranges from approximately 0 to 15 Hz as the HCH angle varies from 120° to 100°. When the separation is three bonds ("vicinal" protons), J varies with the dihedral angle HCH (obtained by looking down the CC bond) approximately as shown in Fig. 3-3. When the separation is more than three bonds,

† Most of the values in these tables are taken with permission from the much more extensive tables in J. R. Dyer, *Applications of Absorption Spectroscopy of Organic Compounds,* Englewood Cliffs, N.J., Prentice-Hall, Inc., 1965, and R. M. Silverstein and G. C. Bassler, *Spectrometric Identification of Organic Compounds,* 2nd Ed., New York, John Wiley & Sons, Inc., 1967. The *Spectra Catalogs* of Varian Associates, Palo Alto, Calif., contain 700 interpreted spectra with an extensive chemical shift index which includes protons in most of the environments commonly encountered in organic compounds. See also *Formula Index to NMR Literature Data,* New York, Plenum Press; this series is unfortunately about 4 years behind the original literature.

Table 3-1

PROTON CHEMICAL SHIFTS

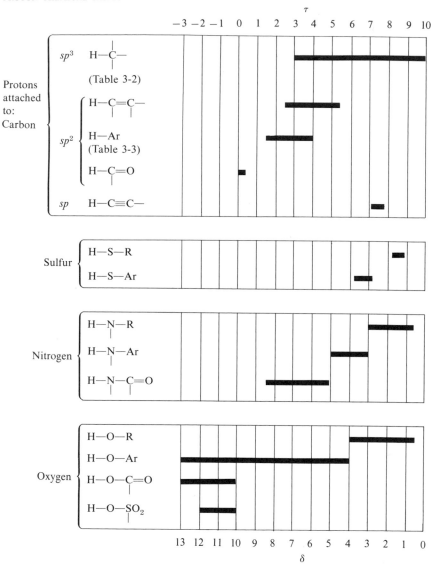

J is usually below 1 Hz and is thus not ordinarily observed except perhaps as a slight broadening of the peaks for the protons involved. This latter effect is observable in Fig. 3-2: The coupling ($J \sim 1$ Hz) between the methyl protons and the vinyl proton causes the vinyl proton absorption to be a broad rather than a sharp triplet. Some coupling constants are summarized in Table 3-4.†

† See footnote regarding Tables 3-1, 3-2, and 3-3 on p. 107.

Table 3-2

CHEMICAL SHIFTS OF PROTONS ATTACHED TO sp^3 CARBON: EFFECT OF α-SUBSTITUENTS

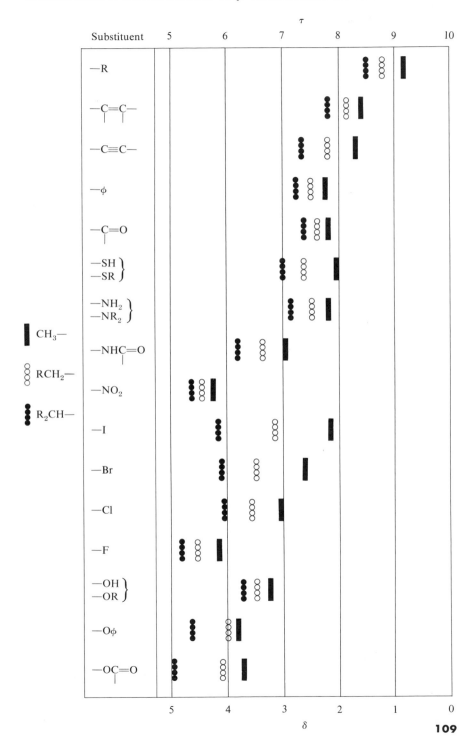

Table 3-3

CHEMICAL SHIFTS OF AROMATIC PROTONS

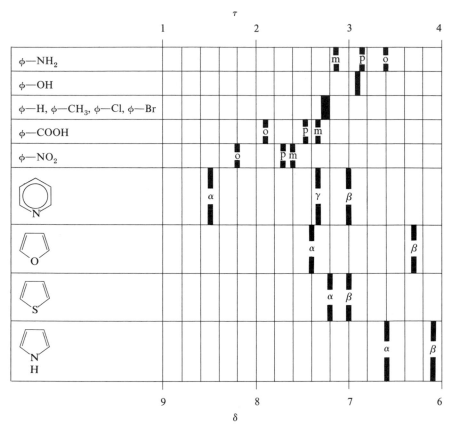

Coupling patterns can be very complex when the coupling constant between two protons is of the same order of magnitude as the difference between their chemical shifts. The patterns are easy to interpret if the chemical shift difference is zero (e.g., protons in a methyl group), since then the spectrum appears exactly as it would if there were no coupling, and also if the chemical shift difference is much larger than the coupling constant. In the latter case, the spectrum shows a splitting of the absorptions of each of the coupled protons into multiplets, with the spacings between adjacent peaks in the multiplet equal to J. The number of peaks in the multiplet and their relative areas are determined by the number of protons causing the splitting and their coupling constants. The simplest case is exemplified by the doublet at about τ 5.9 in Fig. 3-2; these peaks are the absorption for the methylene group in γ,γ-dimethylallyl alcohol, and they occur as a doublet ($J = 6$ Hz) due to splitting by the vinyl proton (joined by three covalent bonds). The vinyl proton, absorbing at about τ 4.5, gives a broad triplet with peak areas 1:2:1, as a result of splitting by two protons (the

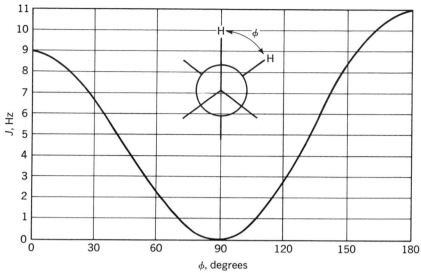

Fig. 3-3 Karplus curve for vicinal protons, from modified Karplus equations
$J = 9 \cos^2 \phi$ ($\phi = 0 - 90°$) and $J = 11 \cos^2 \phi$ ($\phi = 90 - 180°$).

Table 3-4
PROTON-PROTON COUPLING CONSTANTS

	J_{ab} in Hz		J_{ab} in Hz
$\overset{H_a}{\underset{H_b}{\diagup}}C\diagdown$	0–20	$\overset{H_a}{\diagdown}C=C\overset{H_b}{\diagup}$	7–12
$\diagup C=C\overset{H_a}{\underset{H_b}{\diagdown}}$	0.5–3	$\overset{H_a}{\diagdown}C=C\overset{\diagup}{\underset{H_b}{\diagdown}}$	13–18
$\underset{H_a}{\overset{\mid}{C}}-\underset{H_b}{\overset{\mid}{C}}$	0–11 (see Fig. 3-3)	$\diagup C=CH_a-CH_b=C\diagdown$	9–13
$\underset{(a)}{CH_3}-\underset{(b)}{CH_2}-$	6.5–7.5	$\overset{H_a}{\underset{}{\bigcirc}}-H_b$	o 6–9 m 1–3 p 0–1
$\underset{(a)}{(CH_3)_2}-CH_b-$	5.5–7.0		
$\diagup C=C\overset{\overset{\mid}{CH_b}-}{\underset{H_a}{\diagdown}}$	0–10	$H_aC\overset{\mid}{=}C\overset{\mid}{-}CH_b\diagdown$	0–3
$O=C\overset{\overset{\mid}{CH_b}-}{\underset{H_a}{\diagdown}}$	1–3	$H_aC{\equiv}C-CH_b\diagdown$	2–3

methylene protons) with identical chemical shifts (6 Hz). One can imagine each of the peaks of the doublet from splitting by one proton being split again by the second proton to give four peaks of equal area, with the two central peaks coinciding to give one peak of twice the area of each of the outside peaks. Thus, spin-spin splitting is helpful since it shows when protons with different chemical shifts are close to one another, and, because of the variation of coupling constants with dihedral angles in geminal and vicinal cases, it can give information regarding stereochemistry and conformation.

Most spectrometers (including the Varian A-60, for which it is a $5000 accessory) have provision for "spin decoupling" ("double resonance" or "NMDR") experiments, in which the sample is simultaneously irradiated at a second radio frequency, equal to that at which a coupled proton(s) is absorbing. This effectively makes the coupling constant between this proton(s) and all others zero, and thus spin decoupling can be used to simplify complex spectra and to verify coupling patterns. In Fig. 3-2, irradiation at the frequency at which the vinyl proton absorbs would cause the methylene proton absorption to collapse into a singlet and vice versa. A related technique which can be accomplished with the same equipment but which gives in many cases the relative signs of coupling constants is termed "spin tickling."†

NMR can also be used to study certain exchange phenomena, one of which is illustrated in Fig. 3-2. Coupling between the hydroxyl proton (τ 5.3) and methylene protons (τ 5.9) is not observed, even though these protons are connected by three covalent bonds. The reason is that the hydroxyl protons (like OH and NH protons in most NMR samples) are exchanging with one another more rapidly than the NMR measurement, a situation which always gives rise to a sharp peak at a chemical shift corresponding to an average environment. In dimethyl sulfoxide, exchange is sufficiently slow that coupling with nearby protons (in this case, the methylene protons) is observed, providing an easy way of distinguishing primary from secondary from tertiary alcohols.

Since the "average magnetic environment" of OH and NH protons is quite dependent on temperature and concentration, the chemical shifts of these protons are affected much more by these variables than are the chemical shifts of CH protons. Thus, one can detect peaks due to OH and NH protons by running spectra at two or more widely different concentrations (this is generally more convenient than varying the temperature). Alternatively, the spectrum can be run in the usual way, and, if the solvent is immiscible with water ($DCCl_3$, CCl_4), the solution can be transferred back to the sample vial from which it came and shaken for 1 min with one drop of D_2O. A spectrum is then obtained on the organic layer and will show a greatly diminished peak for any OH or NH protons, probably at a some-

† R. Freeman and W. A. Anderson, *J. Chem. Phys.*, **37**, 2053 (1962).

what different chemical shift. After this D_2O treatment, the sample whose spectrum is shown in Fig. 3-2 gives a spectrum identical except that the peak at τ 5.3 is much reduced and slightly shifted.

Another example of an exchange phenomenon conveniently studied by NMR is provided by bullvalene, ⬡, which at room temperature gives complex absorption, but at 100° (using a variable temperature probe) it gives just one peak, due to rapid interconversion of all four types of protons via Cope rearrangements. From the "coalescence temperature" (a temperature at which the spectrum is intermediate in appearance between the two extremes), it is possible to estimate the rate constant for such an exchange process.†

The spectrum shown in Fig. 3-2 illustrates the power of NMR for the derivation of structural formulas from molecular formulas. Given the spectrum in Fig. 3-2 and the molecular formula, an organic chemist with some experience in interpreting NMR spectra could say with confidence after a few minutes of study that the structural formula shown on the spectrum is, of the dozens of possibilities with this molecular formula, the only one compatible with the spectrum. This would not be possible by any other spectral method. The reasoning might go as follows: The peak at τ 4.6 must be due to a vinyl proton from its chemical shift, and since it is split into a triplet with $J = 6$ Hz, there must be an adjacent methylene group. This methylene group, because it absorbs so far downfield, must be attached by a single bond to oxygen. Since there are no further vinyl hydrogens, the remaining two carbons must be in methyl groups attached to the double bond, and the structure is fully defined. The remaining three peaks in the spectrum are entirely consistent with this structure. Several available books have excellent problems which provide invaluable practice in the interpretation of NMR and related spectra.‡

3.3.2 IR Spectrometry: Although the spectrum obtained by plotting the absorbance of infrared (IR) radiation of varying wavelengths by an organic unknown against the wavelength (e.g., Fig. 3-4) usually gives less structural information than its NMR spectrum, the information is complementary and it is helpful to have both spectra when characterizing an unknown. IR spectrophotometers cost less ($3,000–15,000), are more widely distributed in organic laboratories, are easier to use, and are more trouble-free. A few milligrams of sample is sufficient for the usual cells, and with special cells

---

† J. A. Pople, W. G. Schneider, and H. J. Bernstein, *High-Resolution Nuclear Magnetic Resonance,* New York, McGraw-Hill Book Co., 1959.
‡ J. R. Dyer, *loc. cit.;* R. M. Silverstein and G. C. Bassler, *loc. cit.;* B. Trost, *Problems in Spectroscopy,* New York, W. A. Benjamin, Inc., 1967; A. Ault, *Problems in Organic Structure Determination,* New York, McGraw-Hill, 1967; A. J. Baker *et al., More Spectroscopic Problems in Organic Chemistry,* London, Heyden & Son, 1967.

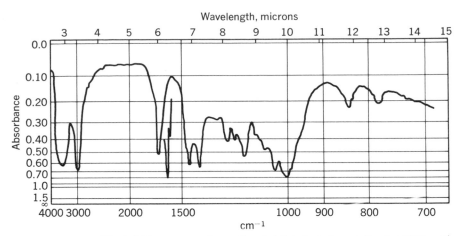

Fig. 3-4 IR spectrum of γ, γ-dimethylallyl alcohol, neat. Band at 3500 cm⁻¹ is OH stretch, at 3000 cm⁻¹ is CH stretch, at 1680 cm⁻¹ is C=C stretch, and at 1603 cm⁻¹ is polystyrene standard.

a good spectrum can be obtained with a few micrograms. The type of cell used will depend on whether the spectrum is run on a solution, a mull or disc (to reduce light-scattering, solids are often mulled with a material such as Nujol or ground with KBr and pressed into a disc),† a pure liquid, or a gas. Gas cells, of course, require the greatest path length (about 10 cm); solution cells are about 0.1 mm thick (for 10% solutions), and pure liquids are usually run as films about 0.01 mm thick prepared simply by putting a droplet of sample between two salt plates and squeezing gently. Cells are usually made of sodium chloride or other inorganic salts because of their near transparency in the region of the infrared spectrum which is usually examined. For solid and liquid unknowns, solution spectra have the advantage that intermolecular effects between solute molecules are minimized, but also the disadvantage that the solvent (CCl_4 or, less often, $HCCl_3$ or CS_2) has some bands in the infrared and thus obscures certain regions of the spectrum. This disadvantage can be overcome by using two or more solvents (usually CCl_4 and CS_2) that do not have any regions of mutual absorption. To minimize the effect of the solvent in running solution spectra, it is common practice with double beam spectrometers to put in the reference beam a solvent-filled cell with slightly shorter path length than the sample cell. Even if the number of solvent molecules in the sample and reference beams is made exactly the same, however, the peaks due to the sample can be grossly distorted in the regions of solvent absorption because of the inability of the detector to distinguish small differences among strong absorptions as well as it can small differences among weak absorptions.

† A simple and very efficient "Mini-press" for preparing KBr pellets can be purchased for $15 from the Wilks Scientific Corp., S. Norwalk, Conn. 06854.

Organic compounds absorb infrared radiation at wavelengths corresponding to differences in *vibrational* energy levels. The information most readily obtained from the IR spectrum is the detection of the presence or absence of certain groupings of atoms whose characteristic absorption band(s) has been empirically determined by examining the spectra of many compounds containing these groupings. Absorptions for some of the groupings whose presence or absence can be reliably determined are given in Fig. 3-5; much more extensive listings are available.† In the figure, peak locations are given in both of the commonly employed units: wavelength, in micrometers ($=10^{-4}$ cm; formerly called "microns"), and *frequency,* in cm^{-1} (reciprocal centimeters or *wave numbers*). The term *frequency* is somewhat misleading since, in its strictest application to electromagnetic radiation, it implies the speed of light divided by the wavelength, not 1 divided by the wavelength. Most spectrometers do not give peak positions very accurately unless special precautions are taken. An easy way to obtain accuracy is to calibrate the spectrum with a peak of known location from the spectrum of a reference compound. Polystyrene is a useful reference compound since it has many sharp bands of known location. It takes only a few seconds to put a polystyrene film in place of the sample in the sample beam and to trace the appropriate polystyrene peak directly on the spectrum of the unknown. The polystyrene peak at 1603 cm^{-1} is particularly useful for this purpose since, besides occurring in a region of great interest, it has a side peak on the low-frequency side whose appearance indicates the resolving power of the instrument at the time the sample is run.

† J. R. Dyer, *loc. cit.;* R. M. Silverstein and G. C. Bassler, *loc. cit.;* L. J. Bellamy, *The Infrared Spectra of Complex Molecules,* 2nd Ed., New York, John Wiley & Sons, Inc., 1958; K. Nakanishi, *Infrared Absorption Spectroscopy—Practical,* San Francisco, Holden-Day, Inc., 1962.

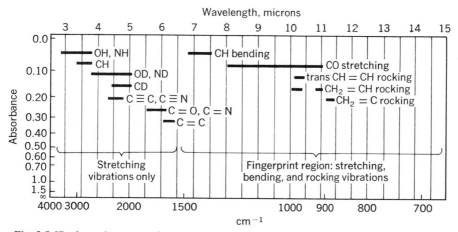

Fig. 3-5 IR absorption ranges for some common groupings.

Peaks in the high-frequency region of the spectrum ($1500-4000$ cm^{-1}) are usually due to simple stretching vibrations and can often be safely assigned to various groupings of atoms, but considerable caution should be exercised in the assignment of bands in the low-frequency region ($600-1500$ cm^{-1}), sometimes called the "fingerprint region" because it contains a large number of peaks useful for establishing identity. Most of the bands in the latter region are associated with more complex vibrational modes. For example, the out-of-plane CH bending mode of monosubstituted aromatics usually gives a strong peak in the $730-770$ cm^{-1} range, but the presence of a strong band in this range is not by itself sufficient evidence for the presence of a phenyl group in an unknown since certain compounds not containing a phenyl group also give a strong band here.

IR spectra would be difficult enough to interpret completely if all bands observed were fundamentals, but unfortunately they are not. Many peaks are due to overtones (the first one occurring at slightly less than twice the frequency of the fundamental), combinations (two vibrations can become excited simultaneously, giving a band at about the sum of their separate frequencies), and Fermi resonance (an insidious phenomenon: Where a fundamental and an overtone or combination band are expected to occur at very nearly the same frequency, one sometimes finds instead two about equally intense bands, one shifted to higher and one to lower frequency).

As suggested in Fig. 3-5, IR provides a valuable means of studying deuteriated compounds. It is a simple matter to calculate the effect that an isotopic change will have on stretching vibrations, using the formula $\nu_{13} = \nu_{23}[M_2 M_1^{-1}(M_1 + M_3)(M_2 + M_3)^{-1}]$, in which ν_{13} and ν_{23} are the vibrational frequencies of the new and old bands, respectively, and M_1, M_2, and M_3 are the masses of the new isotope, the old isotope, and the atom to which these isotopes are attached.

The intensity of an IR band depends on the magnitude of the change in the dipole moment of the molecule during the vibration. If there is no change, e.g., in the double bond stretching mode of ethylene (the stretched and unstretched forms have zero dipole moment), no band is observed; if the change is large, e.g., in the double bond stretching mode of formaldehyde (the more stretched forms have larger dipole moments than less stretched forms), a strong band is observed. Thus, even though there might be 25 similar carbon-hydrogen bonds and only one carbon-oxygen bond in a ketone, the carbon-oxygen stretching vibration, because of the much larger change in dipole moment associated with it, will probably give the strongest peak in the IR spectrum.

3.3.3 Mass Spectrometry: If a mass spectrometer is available, a mass spectrum will probably have been run at an early stage to determine the exact molecular weight and the molecular formula as described above. An attempt may also be made to gain further structural information from the cracking pattern in the spectrum (Fig. 3-1). Since the energy available when an electron collides with a molecule in a mass spectrometer is often sufficient to break many bonds in the molecule, a multitude of different cleavages occur. The details of the processes are usually obscure, but considerable valuable structural information has been obtained by concentrating on the most intense peaks at high mass number. These peaks result from the most likely cleavages of the smallest number of bonds in the molecular ion and are, therefore, the most fruitful to try to interpret. The reactions that occur bear some resemblance to the usual organic reactions, and a familiarity with organic reactions is helpful in interpreting mass spectral cracking patterns. There are differences, however, and it is important to take advantage of previous empirical findings regarding cracking of various classes of compounds.†

3.3.4 UV and Visible Spectrometry: Virtually all organic compounds give NMR, IR, and mass spectral peaks of sufficient utility in structure determination to warrant the small expenditure of sample and time necessary to obtain all of these spectra routinely, provided the spectrometers are available. It is not as generally helpful, however, to obtain an ultraviolet (UV) and visible spectrum, even though all organic compounds show some absorption in this region (100–800 nm). The main reason is that many compounds absorb only in the "far ultraviolet" region (100–200 nm), and since all substances (including air) absorb in this region, a special vacuum spectrometer ($20,000–40,000) must be used to obtain spectra in this little-studied region. The usual spectrometer ($2,000–20,000) covers the near UV (200–400 nm) and visible (400–800 nm) regions.

Absorption of light in the 100–800 nm region causes a molecule to go into an electronically excited state. The usual order of decreasing ease of excitation for various electrons is n (unshared) $> \pi$ (in π bonds) $> \sigma$ (in σ bonds), and only certain compounds having n and/or π electrons are excited at sufficiently low energies to absorb in the near ultraviolet or visible region. Some of the common chromophores (absorbing groupings) with the locations and intensities of their absorption maxima are shown in Table 3-5; much more extensive listings are available.‡ Peak locations are commonly expressed in units of wavelength (λ), either nm (10^{-7} cm; formerly called millimicrons, mμ) or Å (10^{-8} cm). Intensities are in ϵ[= molar absorptivity = optical density/(concentration in moles per liter $\times$ cell thickness in centimeters)], log ϵ, or, if the molecular weight is not known, $E_{1cm}^{1\%}$ (defined like ϵ except concentration in grams per 100 ml). The spectrum is usually

---

† For summaries of common fragmentation paths for various classes of compounds see F. W. McLafferty, *Interpretation of Mass Spectra,* New York, W. A. Benjamin, Inc., 1966.
‡ The values in this table are taken with permission from the more extensive tables in Dyer and Silverstein and Bassler. All UV spectra in the literature are being tabulated in *Organic Electronic Spectral Data,* New York, Interscience Publishers; these volumes are about 4 years behind the original literature. See also A. E. Gillam and E. S. Stern, *An Introduction to Electronic Absorption Spectroscopy,* 2nd Ed., London, E. Arnold, 1959.

Table 3-5

UV ABSORPTION VALUES FOR SOME COMMON CHROMOPHORES

Chromophore	Example	λ_{max}, nm	ϵ
C=C	$CH_2=CH_2$	193	10,000
	$CH_2=CHCH=CH_2$	217	20,900
C=C—C=C (*transoid*)		235	20,400
C=C—C=C (*cisoid*)		265	4,180
C=C—C=C—C=C (*transoid*)	$H+(CH=CH)_3 H$	258	35,000
	C_6H_6	198	8,000
		255	230
C=C		244	12,000
	$\phi CH=CH_2$	282	450
C=O	CH_3COCH_3	279	15
C=C—C=O	$CH_2=CHCOCH_3$	212.5	7,100
		320	27
C=O		240	13,000
	$\phi COCH_3$	278	1,100
		319	50

run on a very dilute solution in 95% ethanol, cyclohexane, or water. For example, a good spectrum can be obtained on 0.1 mg of a conjugated diene with $\epsilon = 10,000$ in a 1-ml cell. Ultraviolet spectra are very useful in showing the presence or absence of certain chromophores, particularly when good model compounds containing these chromophores are available. The shifts to longer wavelengths (*bathochromic* or "red" shifts) and higher intensities (*hyperchromic* shifts) which accompany increasing conjugation are exemplified by the first, second, and fifth compounds in Table 3-5. Increased alkyl substitution is evidenced by small bathochromic shifts (compare the second and third compounds in the table). The absorption is often highly sensitive to the spatial arrangement of conjugated groupings, and use is made of this in structure determination. For example, the wavelengths and intensities observed for the third and fourth compounds in the table indicate that the former is *transoid* and the latter *cisoid*. Woodward's rules for predicting the wavelengths of UV maxima in conjugated alkenes and ketones are valuable for cases in which the spatial arrangement of the groupings in the vicinity of the chromophore is normal.†

3.3.5 Qualitative and Quantitative Microanalysis for Functional Groups: In qualitative organic analysis courses a number of rapid tests for various

---

† L. F. Fieser and M. Fieser, *Steroids,* p. 15, New York, Reinhold Publishing Corp., 1959.

functional groups are learned, and it is occasionally desirable to employ some of these to complement spectral results. Quantitative microanalyses for some functional groups can be run in the laboratory by the organic chemist but more often are performed by commercial laboratories for about $10 per determination (5 mg sample). For example, one commercial laboratory† will analyze quantitatively for carboxyl, methoxyl, ethoxyl, N-methyl, N-ethyl, acetyl, benzoyl, hydroxyl, C-methyl, primary amino, active hydrogen, mercaptan, disulfide, and lactone groupings and will measure the neutral equivalent, saponification equivalent, hydrogen uptake, and iodine number.

3.4 OTHER RAPID METHODS FOR STRUCTURAL
FORMULA, STEREOFORMULA, AND CONFORMATION

3.4.1 Thermodynamic Properties: There are certain situations in which the best way to show a particular structural detail—usually a stereochemical one—is to establish the thermodynamic stability relationship between two or more compounds. To be useful, it is necessary that the stability difference be fairly large, readily measurable (often through direct equilibration), and qualitatively predictable. For example, the best way to determine the stereochemistry of a racemic 2,5-dialkylcyclohexanone would be to equilibrate with base: If it is the major equilibration product, it is the *trans* isomer (which has a diequatorial conformation and is thus more stable); if it is the minor product, it is *cis*.

In cases in which direct equilibration is impossible, it is sometimes possible to use readily measured physical properties which correlate with thermodynamic stability. According to Allinger's "conformational rule," the more stable of two isomers not differing appreciably in dipole moment is usually the one with the lower density, refractive index,‡ and boiling point. There is some indication that the first two of these correlate more reliably than the last.§

3.4.2 Dipole Moments: While dipole moments are not generally useful with organic unknowns, occasionally they provide the best means of distinguishing among a few possible structures which would have differing dipole moments. Approximate moments can be calculated by vector addition of bond moments from the literature.¶ For example, *trans*-1,4-dibro-

---

† Huffman Laboratories, Inc., Box 350, Wheatridge, Colo. 80033.

‡ Measured on a droplet of a liquid with a refractometer ($130–2000) and referred to with the symbol n_D^{25}, for example, in which the superscript is the temperature in degrees centigrade, and the subscript indicates that the measurement is made at the D line of sodium. For many organic liquids, the refractive index is inversely proportional to the temperature with the proportionality constant about 0.0004 per degree.

§ E. L. Eliel, *Stereochemistry of Carbon Compounds*, p. 327, New York, McGraw-Hill Book Co., 1962.

¶ E. A. Braude and F. C. Nachod, *Determination of Organic Structures by Physical Methods*, Vol. I, p. 390, New York, Academic Press, 1955.

mocyclohexane can be distinguished from the *cis* isomer by virtue of the large difference in dipole moments between the former (zero dipole moment since centrosymmetric) and latter (dipole moment, several Debyes). Even without measuring the dipole moments of this pair of stereoisomers, it would have been possible to make at least tentative configurational assignments from measurements of density, refractive index, or boiling point. These assignments would be based on the "dipole rule," which states that for stereoisomers differing considerably in dipole moment the isomer with the greater dipole moment will also have the greater density, refractive index, and boiling point.†

Experimentally, the dielectric constant and density of the substance are measured. Either the dielectric constant is measured at several temperatures or the refractive index must be measured; the dipole moment is then calculated. The dielectric constant is measured in a conductance cell on the vapor or, more often, in dilute solution in a nonpolar solvent such as benzene.‡ For accurate work, the dielectric constant is measured at several concentrations to allow extrapolation to infinite dilution.§

3.4.3 Optical Rotatory Dispersion:¶ With optically active compounds, it has been common practice for many years to measure the optical rotation at 489 nm, the wavelength of the sodium D line, using a polarimeter (typical cost, $500).

The equation $[\alpha]_D = \alpha/lc$, in which α is the observed rotation in degrees, l is the length of the sample tube in decimeters, and c is the concentration of sample in grams per 100 ml, is used to calculate the specific rotation $[\alpha]_D$. The molecular rotation can be calculated (when the molecular weight of the sample is known) by multiplying the specific rotation by the molecular weight.

Such measurements at one wavelength are now being replaced by more revealing plots of rotation versus wavelength over the wavelength range 250–700 nm called *optical rotatory dispersion* (ORD) curves. The curve is particularly revealing if there is a chromophore present (such as a saturated ketone grouping) which absorbs weakly in this region, since there will then be a maximum and minimum in the curve, the pattern of which, with the aid of suitable model compounds or the "octant rule," can yield configurations and conformations in the vicinity of the chromophoric group. ORD instruments, usually called *spectropolarimeters,* cost $10,000–35,000. Some instruments give as output a curve showing the difference between absorp-

---

† E. L. Eliel, *Stereochemistry of Carbon Compounds,* p. 327, New York, McGraw-Hill Book Co., 1962.

‡ Kahlsico, Box 1166, El Cajon, Calif. 92022, sells a dipole apparatus for $1500.

§ The dipole moments recorded in the literature up to 1961 have been compiled in A. L. McClellan, *Tables of Experimental Dipole Moments,* New York, W. H. Freeman & Company, 1963.

¶ C. Djerassi, *Optical Rotatory Dispersion,* New York, McGraw-Hill Book Co., 1960.

tion for right- and left-hand circularly polarized light; this technique is called *circular dichroism*. The sample is run in solution, with the concentration depending on the path length and the rotation of the sample. The usual sample size is several milligrams, with more being necessary if the specific rotation is very low and with less if it is very high.

3.4.4 ESR Spectrometry:† One of the most powerful techniques for the study of substances with one or more unpaired electrons makes use of electron spin (or paramagnetic) resonance, ESR (EPR). The spectrometer ($20,000–35,000) bears some resemblance to an NMR spectrometer. This is not surprising since NMR and ESR both deal with the absorption of electromagnetic radiation by particles in a strong magnetic field, the former by nuclei (usually protons) and the latter by unpaired electrons. The spectrum is usually a plot of the first derivative of absorption versus magnetic field strength. Most organic unknowns give no signal whatsoever, but those which do contain unpaired electrons give such strong signals that they can be detected in concentrations as low as 10^{-12} M. The sample may be run as a solid, liquid, or gas. The fine structure of the absorption, caused by spin-spin splitting by nearby magnetically active nuclei (usually protons), makes the ESR spectrum useful as a fingerprint for various radicals and, if the coupling pattern can be deciphered, also yields structural information for the part of the radical in the vicinity of the unpaired electron.

3.4.5 Raman Spectrometry: When molecules scatter monochromatic ultraviolet and visible radiation, in addition to scattered light of the same frequency as the incident light, some scattering at the incident frequency plus and minus certain molecular vibrational frequencies occurs. This latter scattering is the Raman effect. Raman spectrometers are commercially available (about $25,000) and give spectra which are occasionally useful in organic structure determination.‡ The selection rules for IR and Raman spectra are quite different. For example, as mentioned above, the carbon-carbon stretching band in ethylene is infrared inactive; it is, however, Raman active. The selection rules are such that a molecule with a center of symmetry will have no bands common to both spectra, and this proves useful in detecting centrosymmetry. The sample is run as a pure liquid or as a solution of at least 5% concentration. The usual cells require about 100 mg of sample, but with microcells the minimum sample size is a few milligrams.

3.4.6 Stability Constants: There are certain cases in which the equilibrium constant for a reaction of the sample with some substance can be easily measured and yields valuable structural information. Useful structural information is forthcoming only when there have been previous studies of related equilibria, as is the case with acids and bases.

† The most comprehensive work on experimental aspects of ESR is C. P. Poole, Jr., *Electron Spin Resonance,* New York, John Wiley & Sons, Inc., 1967.
‡ For example, see G. A. Olah, A. Commeyras, and C. Y. Lui, *J. Am. Chem. Soc.,* **90**, 3884 (1968); this paper reports a laser Raman study of a carbonium ion.

A revealing and characteristic property of any acid or base is its ionization constant, K_a, which is usually expressed in terms of its negative logarithm, pK_a. A comprehensive analysis of pK_a values often can provide quantitative information on the magnitude of steric and electronic interactions in addition to other varied details, such as the conformation of an acidic or basic group or the site of protonation of an amino acid. The precise determination of a pK_a value is not difficult, but it does require careful technique if meaningful numbers are to be obtained. The analytical method of choice is dictated by the range of the pK_a value of the substance and its chemical and physical properties; but only potentiometric titration, the technique that is used for most organic acids and bases, is discussed here.

The necessary equipment is a good pH meter, a burette, and a jacketed titration cell (Fig. 3-6). The carefully weighed sample is added to the titration cell and a known volume of pure solvent is added, the mixture is stirred magnetically until the acid (or base) has dissolved completely, and a slow stream of nitrogen is passed through the solution. The glass electrode is then standardized in a buffer solution and placed in the cell. When the solution has come to thermal equilibrium with the cell, the addition of increments of carbonate-free base is begun and the pH is recorded as soon as equilibrium is attained after each addition. The exact value of the pK_a' should be

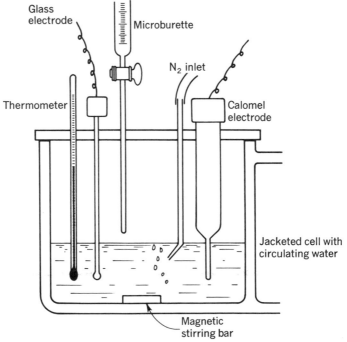

Fig. 3-6 Apparatus for pK_a determination.

calculated using all the data obtained, but it may be estimated from the pH value at half neutralization.†

3.4.7 Kinetics: One of the many applications of studies of reaction kinetics (see Sec. 1.7.3 for experimental details) is in the determination of structural formulas and especially stereoformulas. A familiar example is osazone formation: Glucose, mannose, and fructose give the same osazone, but at quite different rates so that they can easily be distinguished by a kinetic study. As another example, suppose a complex cyclohexanone is reduced with lithium aluminum hydride to give a mixture of two cyclohexanols. If one of these alcohols is acetylated much more rapidly than the other, it is very likely that the former has the equatorial hydroxyl group and the latter the axial hydroxyl group. Often qualitative observations of relative rates will serve as in this case.

3.5 MORE TIME-CONSUMING METHODS FOR STRUCTURAL FORMULA, STEREOFORMULA, AND CONFORMATION

If application of an appropriate selection of the rapid methods above has not given the structure in the desired detail and with the desired degree of certainty, one or more of the following more time-consuming methods may be employed to complete the structure determination.

3.5.1 Degradation and Synthesis: These chemical methods, for many years the mainstay of structure determination, are not *always* very time-consuming, depending on how many essentially unambiguous reactions are necessary to convert the unknown into one or more compounds of known structure or vice versa. As an extreme example, the acetate of a triterpene alcohol of known structure might be found to occur in nature; it would best be characterized by identifying its saponification products as acetic acid and the triterpene alcohol and/or by showing that the acetylation product of the triterpene alcohol is identical with the natural acetate. On the other hand, dozens of man-years have been required in many cases to learn one complex structure by these methods alone. The most complex structural problems are now usually solved by diffraction methods (next section), but most structures are still determined by a combination of degradation and synthesis with the rapid methods described above.

It is very important in a degradation or synthesis used in structure determination to try to employ reactions whose course is highly predictable. In particular, reactions involving carbonium ion intermediates should be avoided whenever possible because of the great ease with which many carbonium ions rearrange. Thus, the best way to replace the hydroxyl group in a secondary alcohol by a hydrogen atom is not to dehydrate with sulfuric

† For a more detailed discussion of this and other methods, see A. Albert and E. P. Serjeant, *Ionization Constants of Acids and Bases,* New York, John Wiley & Sons, Inc., 1962.

acid (carbonium ion intermediate) and hydrogenate but to run a Wolff-Kishner reduction on the corresponding ketone or to reduce the tosylate with lithium aluminum hydride. Dehydrogenations, zinc dust distillations, and alkali fusions frequently (and not very predictably) proceed with rearrangement, and while they are often very useful in the early stages of a structure determination for gaining some information regarding gross structural features, it is highly desirable to find eventually one (or more) line of evidence independent of such reactions.

Synthesis is not used, of course, until the likely structural possibilities have been limited to one or a few. It is most often used for final verification of a quite well-established formula, but sometimes the easiest way of distinguishing between two equally likely formulas is to synthesize one or both.

3.5.2 Diffraction Methods: X-ray diffraction, the most widely applied diffraction method, is applied to crystals and gives not only all stereochemical details but the conformation in which the substance crystallizes, the packing arrangement in the crystal, intra- and intermolecular bond distances and angles, and the vibrational motions in the crystal. The option of operation well below room temperature cancels in part the disadvantage that the sample must be crystalline; still, there are many substances of interest that do not give sufficiently good crystals for this method. The major disadvantage—the several months required for one structure determination—is being diminished with automation of data collection, faster computers, and new and improved methods for the solution of the phase problem.

When X-rays pass through a crystal, a certain amount of scattering occurs, yielding a characteristic pattern of spots that can be detected with a photographic plate or a scintillation counter. While the calculation of the scattering pattern from the structure of the scattering material is straightforward (with access to a digital computer and standard programs), the more important problem of calculating the structure from the scattering pattern is not. The origin of the difficulty, called the *phase problem,* is not considered here, but it is noted that this is readily solved for most optically inactive substances and for optically active substances containing a few heavy atoms, i.e., atoms with atomic weight greater than *ca.* 20.

A single crystal 0.1–1 mm on a side (weight about 0.1 mg) is mounted on a goniometer head and several pictures are taken using a Weissenberg or precession camera (*ca.* $3000). This preliminary examination reveals the dimensions of the unit cell and also (usually) in which of the 230 space groups the compound has crystallized. The latter finding allows one to take advantage of the symmetry in the space group, greatly simplifying the later calculations. The second step consists of indexing spots (perhaps 2000 of them) according to the crystal plane which gives rise to them and measuring their relative intensities. This is sometimes done by taking another series of Weissenberg and/or precession photographs and estimating the intensities visually. For greater accuracy (but with a loss of time) a densitometer

($1,000–10,000) may be used to measure the intensities. Alternatively, an automatic diffractometer ($30,000–100,000), that employs a scintillation or proportional counter rather than a camera, may be used to obtain an even more accurate diffraction record in a shorter time.

The last step consists of processing the intensity data with the aid of a digital computer to derive the structure. A good test of the correctness of a structure is available: If an approximate structure "refines" satisfactorily (using a refinement program), the correct structure results. The progress of a refinement is judged by the behavior of the "R factor," which is approximately the average percent difference between observed and calculated intensities. R factors below 15% are usually satisfactory. The most difficult part (due to the "phase problem") is to find a sufficiently close trial structure for refinement. This is easy for compounds containing one or two relatively heavy atoms (e.g., bromine), and for this reason heavy atom derivatives are often prepared for crystal structure analysis. When no heavy atom is present, "direct methods" are often successfully employed. In many cases, considerable ingenuity is required in the interpretation of preliminary crude electron density maps from which the first few atomic positions must be assigned.†

Electron diffraction and neutron diffraction have been much less widely applied and are usually used when X-ray diffraction is not suitable, especially electron diffraction for liquids and gases and neutron diffraction to locate hydrogen atoms accurately.

---

† The best book for a beginning crystallographer is G. H. Stout and L. H. Jensen, *X-ray Structure Determination,* New York, The Macmillan Co., 1968. He will also find indispensible the *International Tables for X-ray Crystallography,* Vols. I–III, Kynoch Press, Birmingham, England, 1952–1962. If X-ray computer programs are not already on hand, he should purchase the *World List of Crystallographic Computer Programs* (updated every few years) from the Polycrystal Book Service, Box 11567, Pittsburgh, Pa. 15238; this booklet lists programs available free of charge for most common computers.